THE LITTLE
BOOK FOR
PLANT
PARENTS

FELICITY HART

summersdale

THE LITTLE BOOK FOR PLANT PARENTS

An Hachette UK Company
www.hachette.co.uk

Summersdale Publishers Ltd
Part of Octopus Publishing Group Limited
Carmelite House
50 Victoria Embankment
LONDON
EC4Y 0DZ
UK

www.summersdale.com

Printed and bound in China

ISBN: 978-1-78783-687-7

Substantial discounts on bulk quantities of Summersdale books are available to corporations, professional associations and other organizations. For details contact general enquiries: telephone: +44 (0) 1243 771107 or email: enquiries@summersdale.com.

CONTENTS

INTRODUCTION

Every plant parent starts out with the best intentions, but sometimes – even with plenty of space, a religious watering schedule and plenty of sunlight – we still fail to keep our fledging flora alive. How many of us have joked about our steadfast talent for killing plants off as fast as we buy them? Well, no more.

This pocket guide won't only show you how to keep your plants alive – it'll provide all the secrets to helping them thrive, and take you on a journey with plenty of surprises along the way. Did you know, for example, that having plants around your home can improve mood, increase positivity, fight stress and even increase your pain tolerance? Plus, filling your home with a special green gang can also reduce blood pressure, headaches and fatigue.

The following pages offer tips for perfect plant care, simple solutions to any bumps along the road, and fun DIY makes for keeping your collection as chic as possible. With plant profiles to provide you with a miniature library of indoor horticulture, you'll always know how to pick the right species to flourish in your home.

PLANT CARE 101

If you've never picked up a trowel before,
this chapter is for you. It covers all the
basics – from choosing the right soil to
cleaning the leaves. Before you know
it, you'll be caring for your plants like
a true green-fingered professional.

HOW TO CHOOSE THE RIGHT PLANT FOR YOU

Choosing your plant babies is one of the most exciting moments of plant parenthood – and there's so much choice! From glossy green giants to shiny squat succulents – the possibilities are practically endless. To avoid being overwhelmed, here are the questions to ask before you make a decision:

How much light does my home get?

It's always useful to research your plant's natural home, to see if you'd be able to provide a similar environment for them in your own space. Cacti, for example, thrive in the desert where bright sunlight is abundant – as such, they'll grow best in the sunniest part of your home. Philodendrons, however, grow underneath tropical canopies, so they're accustomed to limited light and can tolerate dark or shaded environments.

How much space do I have?

What are you hoping to achieve from your houseplant collection? A few plants to accentuate a small space, or a full-on indoor jungle? Some plants may begin life as cute little greens, but soon shoot up to become leafy goliaths. You'll want to ensure you'll have the room to accommodate them way past their first baby shoots. If you're lacking in floor space, hanging plants are a great choice.

What's the humidity like?

Though some plants thrive in humid environments, for others it's the seal of death. Be sure to factor the humidity of your space into your choice of plant.

Will my pets love it too?

What might be a friendly frond to you can be a poisonous enemy to your furry housemates. If you have pets, always check that your choice of plant won't be dangerous for them before you purchase. Common favourites such as snake plants, pothos, peace lilies and philodendrons can all be toxic to animals.

CHOOSING THE RIGHT SOIL

All good plant parents want their fronds to feel at home in their soil. Here's what you should look out for when choosing your blend:

1   Avoid soils that include pine bark, as this can attract gnats to your plant.

2   Mixtures that include limestone are an especially good choice, as they balance out the acidity of your soil and provide the perfect pH for your plant to flourish.

3   If you're nurturing cacti or succulents, choose soil with high water retention; this will prevent their roots from rotting or becoming oversaturated. Peat moss and perlite are great water-retaining ingredients to look out for.

4   If your plants are top-heavy, mix sand into your soil to give strength to their base and help to drain away excess water.

When in doubt, choose a drier soil than you think, as most plants would rather be too dry than too damp. You can add more water, but you can't reverse the damage of root rot.

HOW TO WATER YOUR PLANT

As a general rule, you can judge how much water a plant needs by the thickness of its leaves. Thin, delicate-leaved species, such as ferns, can dry out quickly, and so will need generous watering. Succulents and cacti, however, have thick leaves/bodies, which store moisture for long periods of time – too much water can drown these plants.

Before you grab the watering can, check your soil. Generally, if your plant needs a drink, the soil will be dry for 5 cm (2 inches) below the surface – not just on top.

If your pot has drainage holes in the base, it's best to place your plant outside or in the sink when watering, to allow liquid to drain. This is much better than letting your plant sit in a wet container, as this can cause rot to set into the roots.

Every plant has different preferences when it comes to hydration, so be sure to develop a care plan to suit their unique needs. It's often best to check the label on each plant you purchase (if possible) to make a note of the suggested watering guidelines.

HOW DO I KNOW IF/WHEN TO REPOT?

Knowing when it's time to repot your plant – and how to do that while causing as little trauma as possible – is key to successful plant parenting. Here are the four indisputable signs that your plant is ready for a new home:

- The soil is drying out fast, and water doesn't seem to soak in anymore.
- New leaves grow slowly and are smaller than before.
- When you pull your plant out of its pot, you can see a lot of root, and very little soil.
- You can see some of your plant's roots poking through the surface of the soil or stretching through the drainage holes of your pot.

The best time of year to repot your houseplant is in the spring, before any new growth starts to accelerate. Here's a basic guide to repotting*:

Lay down a plastic sheet or some newspaper – this is going to get messy!

1   Carefully upend your plant to remove it from its container. If it's particularly overgrown, you may need to deliver a sharp thump to the bottom of the pot or ease it around the sides with a kitchen knife.

2   Place a little soil in the bottom of the new pot. Gently lower the plant in, making sure it sits at the same height as before. Fill any gaps with new compost.

3   Press the compost down and give your plant some water, to moisten its new soil. Et voila! A lovely new home.

TIP: If you're tired of making a muddy mess every time you repot your plants, try using an ice cream scoop in place of a trowel. It's the perfect shape to prevent spillage.

* If you've recently bought a plant, give it a week to acclimatize to your home before repotting it – otherwise they may go into shock and shed their leaves.

CLEANING YOUR PLANT

When houseplants become dusty, they don't get as much light; this interrupts photosynthesis, which is vital to their nourishment. Although a spritzing bottle is a great way to prevent dirt build-up, sometimes your plants need a deeper cleanse.

There are two ways to clean your plant, depending on its fragility. For plants with fine, delicate leaves, dampen a cloth with warm (not hot) water, and gently rub its leaves in circular motions. This is also brilliant for removing pests (particularly spider mites and aphids). To prevent disease from striking, finish your plant's make-over with a haircut, removing any leaves or stems that have died or turned brown.

If your plants are after a more intense dose of TLC, stir up a mixture of milk and water, and gently rub the solution over the leaves with a soft cloth. This will leave a glossy finish. Repeat whenever you notice an excessive build-up of dust.

TROUBLE-SHOOTING

It's every plant parent's worst nightmare:
Your leaf-baby is green and thriving
until, all of a sudden, it starts to wilt,
turn yellow or attract flies. Fear not!
No matter how mysterious your plant's
maladies seem, our trouble-shooter will
help you to nurse them back to health.

WHAT TO DO IF YOUR PLANT HAS BUGS

Bugs are any houseplant devotee's worst enemies. Not only are they bad for your plant's health, but they can also create a mess – and nobody wants extra creepy crawlies in their home. Here are the best quick fixes for an unwanted infestation:

Aphids

Aphids are nearly invisible to the naked eye, but you can detect their presence through the havoc they wreak; look out for tiny, sticky holes or yellow/misshapen leaves. To rid your plant of these pests, combine equal parts of rubbing alcohol and water with a drop of dishwashing detergent, and apply to the entire plant using a soft brush or cloth.

Spider mites

Hold your plant over a white piece of paper and gently shake. If small, dark specks fall onto the sheet, it's likely you have spider mites. If you suspect one of your plants has been infested, quarantine it immediately, as the mites can easily spread.

Spider mites love dry warmth, so a quick tepid shower can often be enough to dissuade them, and frequent spritzing will discourage their return. For more persistent cases, create the following solution and regularly mist over your plant:

- 120 g (4 oz) plain flour
- 60 ml (2 fl oz) buttermilk
- 4 litres (1 gallon) cool water

Fungus gnats

Are any pests more annoying than the tiny, buzzy flies that gather around your favourite houseplant? Unfortunately, swatting the occasional offender isn't enough to rid your plant of gnats long-term. Instead, make it your mission to prevent the top soil of your plant from becoming moist through overwatering, which is what attracts them in the first place.

WHAT TO DO ABOUT LEAVES DROPPING OFF

It's normal for a plant to shed leaves every now and then, but if the rate of leaf loss is leaving your plant patchy, it could be a sign that something is wrong.

Plants often react to a lack of water by shedding their leaves. This is because fewer leaves mean the plant has less of itself to sustain, and so it can prolong its lifespan. If you frequently forget to water your plant, try this simple trick: fill a lidless, empty bottle with water, turn it upside down and push into the soil. This will provide a gentle, steady stream of hydration. Use water or wine bottles for large plants, and spirit miniatures for smaller plants.

If you've ruled out underwatering, your plant may have suffered a shock. If you've recently changed its environment by moving or repotting it, then stress could be causing it to drop its leaves. There's no quick fix for a shocked plant but feeding it with a weak solution of sugar and water has been shown to reduce recovery time.

A CUT ABOVE

Pruning your favourite houseplants might be painful (after working so hard to help them grow), but a regular trim is as good for your plant as it is for your hair. If your plant is looking unbalanced or growing too tall for your space, it's time to break out the shears. Here's how to give your houseplant a haircut:

1   If in doubt, don't cut it out. It can take a long time for your plant to regrow once it's been trimmed, so only cut areas that appear dead or are becoming too big for your space. As a rule of (green) thumb, never prune more than 25 per cent of your plant at any one time.

2   Using sharp shears or kitchen scissors, slowly begin to prune, stepping back regularly to ensure you keep a balanced view of the entire plant. When cutting back branches, always stop just above the bump on the stem where new growth will appear. If you're removing a frond completely, cut as close to the stem of the main plant as possible.

3   If your plant is on the delicate side, you can choose to "pinch" their growth instead. To do this, pinch their stems between your thumb and forefinger nail to detach them from the plant. This will ensure bushy regrowth and prevent them from becoming "leggy" (excessively long).

4   If you hate to let your clippings go to waste, don't worry – most cuttings can be propagated into new plants. Place them in a cup of water and plant in soil once the roots start to set in. You should have new growth in just a few weeks.

TIP: Pruning your plant leaves it vulnerable to disease. Clean any cutting instruments before use in a mild bleach and water solution. And, a word of warning: never prune palms, as cutting the wrong bud can kill the entire plant.

WHAT TO DO ABOUT DROOPING LEAVES

Drooping leaves are a sorry sight on a formerly perky houseplant. And it's not surprising they look so sad – wilted leaves are a sign that your plant isn't getting enough water. Without adequate hydration plants can't photosynthesize, which means they can't make any food. We'd be looking pretty dejected, too! Water your plant more often and you should soon see them return to their sprightly self.

If your plant is still drooping despite regular watering, try placing a few ice cubes on the surface of its soil (but take care that they aren't actually touching your plant). As the ice cubes slowly melt, they will provide hydration throughout the day.

The next time you repot any persistent droopers, place a damp sponge into the bottom of their new home before filling it with soil. The sponge will act as a water reservoir, delivering a much-needed drink when your plant begins to slouch.

STYLING YOUR PLANTS

So, you've chosen your plants, kept them alive and beaten off any ailments – now the real fun can begin! Whether you want to DIY a stylish hanging planter, get crafty making a bespoke pot, or learn how to take the perfect #plantspo pic, the following pages will help make your plant collection utterly unique.

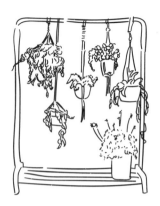

NO-NAILS HANGING PLANTS

If you can't drill holes or add hooks to your property, a repurposed clothing rail is a fantastic way to add some hanging plants to a room – without losing your deposit. Plus, it's easy to move around, creating an elegant impromptu partition to any room, and allowing you to reimagine your space whenever you please.

You'll need:

- A basic clothing rail
- Macramé planters (see p.26 to DIY your own)
- Rope for fastening
- Spray paint (optional)

Method:

Step 1: If desired, use your spray paint to change the colour of your clothing rail. Be sure to spray in a well-ventilated area (ideally outside).

Step 2: Once the spray paint is completely dry, fasten your macramé planters to the clothes rail using the rope. Hanging your planters at different heights will create a more stylish display.

Step 3: Place your plants into the macramé hangers and enjoy!

No space for a clothes rack? Hanging plants onto your shower rail is another great hack for renters.

EASY MACRAMÉ PLANT HANGER

This classic planter design is far easier to create than it looks and is best suited for planters of between 13 and 18 cm (5 and 7 inches) in diameter.

You'll need:

- Scissors
- Long cotton rope
- A 5 cm (2 inch) wooden ring

Method:

Step 1: Cut three pieces of rope so that they measure 1.4 m (55 inches) each.

Step 2: Fold all three pieces of rope in half. Take the mid-point of the rope (the "loop") and pull it a little way through the ring. Then take the loose ends of rope and pull them through the loop. Tighten the knot against the ring.

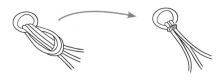

Step 3: Starting 20 cm (8 inches) down from the ring, take two loose ends, tie a half knot, and then tie another half knot in the opposite direction. Pull from each side to secure the knot. Repeat twice more with the four remaining loose ends.

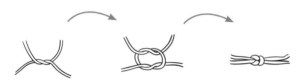

Step 4: 8 cm (3 inches) down from your row of knots, tie another series of three.

Step 5: 10 cm (4 inches) down from your second series of knots, gather all six ropes together and tie into one large knot.

Step 6: Trim the rope and hang from ceiling hook.

COCONUT HANGING PLANT HOLDER

Upcycling is the name of the game with this cute hanging plant hack.

You'll need:

- Half a coconut shell
- Acrylic paints
- Hand drill
- 3 m (10 ft) length of string, cut into four
- A keyring hoop

Method:

Step 1: Paint your coconut with your chosen design and leave to dry.

Step 2: Using your drill, add five small holes to the coconut, two on each side (about 0.5 cm (1/5 inch) below the rim) and one in the bottom for drainage.

Step 3: Thread a piece of the string through each of the top holes.

Step 4: Add your soil and plants to the coconut shell.

Step 5: Gather the tops of the string, tie them to the hoop and hang the plant from a ceiling hook.

WOODEN BOWL PLANT STAND

Repurpose a plain fruit bowl into a chic plant stand with this simple DIY.

You'll need:

- Acrylic paints (optional)
- A flat-bottomed wooden bowl
- Three table legs (purchasing cheap second-hand tables and removing the legs is a great way to source these)
- Screwdriver and screws

Method:

Step 1: Use acrylic paints to add a design onto your wooden bowl if desired.

Step 2: Attach the table legs to the bottom of your bowl, using your screwdriver and screws.

Step 3: Place your plant in the bowl and enjoy!

BRIGHT TIN PLANT POTS

This DIY brings a whole new meaning to the phrase, "one man's trash is another man's treasure."

You'll need:

- An empty, clean tin (be careful of sharp edges)
- Acrylic paints in your chosen colours
- Permanent marker (optional)

Method:

Step 1: Paint your tin in the base colour of your choice and leave to dry.

Step 2: Paint your desired pattern over your base colour; bright pastels tend to work well over a white background. If you're unsure about your design at first, mark out your idea using pencil (you can always come back and erase the pencil marks later).

Step 3: Use a permanent marker to add extra detail or definition to your design. Metallic permanent markers offer an added shimmer.

Step 4: Allow to dry and add your houseplant.

GO FOR GOLD

These simple-yet-luxurious pots make great gifts, as well as stylish additions to your own plant collection.

You'll need:

- A terracotta pot
- Glue
- Gold or silver leaf (or imitation gold or silver leaf)
- Spray sealer
- Gloves

Method:

Step 1: Brush a leaf's length of glue onto the pot.

Step 2: Instead of waiting for the glue to dry (terracotta is porous, and the glue will dry out before it goes tacky) adhere the leaf to the pot immediately.

Step 3: Continue this process around the rim of the pot until covered to your satisfaction.

Step 4: Cover the pot in the sealing spray and leave to dry.

TIP: Do not paint over any gold leaf you've already applied, as this will mute the shine. You may also wish to wear gloves, to avoid getting any glue on your skin.

MOSAIC PLANT POTS

Patience is key in creating these intricate, crafted pots.

You'll need:

- Ceramic tiles
- A hammer
- Wooden spatula
- Tile adhesive
- Plant pot
- Round decorative craft mirrors
- Grout powder
- Craft paint

Method:

Step 1: Make your mosaic tiles by placing the ceramic items in a thick bag/newspaper and breaking them up with a hammer.

Step 2: Use a wooden spatula to spread an even layer (around 0.5 cm (0.2 inches) thick) of your adhesive glue over the area of the plant pot that you wish to embellish.

Step 3: Press your small tiles and decorative mirrors onto the adhesive, leaving small even spaces in between the tiles (you will grout these later).

Step 4: Once covered, leave to dry for at least 3 hours.

Step 5: Create your grout according to the packet instructions. Mix in a little craft paint in your chosen colour, then spread over the pot. Don't worry about overlap on the tiles as you can wipe this away later.

Step 6: Allow grout to dry for 2 hours, then wipe the pot down with a damp sponge. Once completely dry, wipe the pot down with a soft cloth.

SAY CHEESE (PLANT)

Every proud plant parent will know the desire to share the fruits of your labour with the (internet) world. If you're planning to post your favourite plant to Instagram, follow our advice on catching their best side.

- Move your plant of choice closer to a window before you take your shot, since natural light will highlight the beauty of their features. Avoid flash or artificial light which will flatten the image.
- A group of plants can look boring if they're all the same height; get creative by mixing the levels using stands or steps.
- Avoid cluttered or distracting backgrounds to ensure your plant is the star of the show.
- Pick your moment: capturing your plant when it's freshly watered, or recently repotted will show them off at their most glossy.
- Get creative! Don't simply copy the #plantgoals posts you see on social media – try to take photographs that showcase your plant's unique personality.

WE'RE ALL IN THIS TOGETHER

Plants benefit greatly from growing in groups. Grouping together plants that all thrive in humid conditions, for example, can help to create a pocket of moisture that will benefit each plant. However, if you get the combination of your plants wrong, the health of your plants may suffer. From an aesthetic standpoint, you want to avoid arrangements that look overly cluttered, clashing or unnatural. Here are three things to consider:

1   **Height.** Position taller plants toward the back of your arrangement and stagger the others so that your very smallest are toward the front. This ensures that each plant is equally on display and draws the eye pleasingly around your arrangement.

2   **Contrast.** Keep your cluster diverse in texture and colour. Plain, deep greens pop brilliantly when placed alongside colourful or heavily patterned plants.

3   **Potted pals.** Ensure your pots complement one another. Pick similar tones or a colour scheme for your cluster to ensure individual planters don't steal the show.

4   **Group by need.** Make sure the plants in your green gang don't have competing needs so you can display them in a spot that suits them all.

PLANT PROFILES

Trying to decide between a peace lily and a palm? Wondering whether you should put your new plant baby in the living room or the bathroom? If so, this chapter is for you – it's full of ideas, handy tips and plenty of plant-spiration.

LIFE IS SHORT;
BUY MORE PLANTS.

IN ALL THINGS OF
NATURE, THERE IS
SOMETHING OF
THE MARVELLOUS.

ARISTOTLE

PLANTS WILL ALWAYS LOVE YOU BACK.

SNAKE PLANT
Dracaena trifasciata

Snake plants are perhaps the most chill of all houseplants. They require little maintenance, are extremely hardy and look structural and stylish. Though almost anybody can keep these plants alive and thriving, only the true houseplant-whisperer can encourage them to flower – it's a labour of love that's well worth the effort. Snake plant flowers are small white delights that look like miniature lilies – a badge of honour for a plant parenting job well done.

- **Watering:** Overwatering is the biggest danger to this hardy plant. Ensure the top 5 cm (2 inches) of soil are dry before you dive in with the watering can.
- **Light:** Snake plants are a little light-shy and prefer spots away from direct sunshine.
- **Size:** Standard snake plants will stand at around 35 cm (14 inches) tall, but they can grow up to a metre (3 feet) in height.
- **Pet friendly?** No. Snake plants are toxic to cats and dogs.
- **Top tip:** Snake plants are prone to gathering dust on their long leaves. Remove the dust regularly using a damp cloth, to ensure your plant continues to grow.

HOME
IS WHERE
YOUR
PLANT IS.

I'M A PRETTY CONTROLLED AND DISCIPLINED PERSON, BUT MY REAL VICE IS BUYING PLANTS.

ZAC POSEN

VENUS FLY TRAP
Dionaea muscipula

Venus fly traps have fascinated us for years with the satisfying snap of their lobes, which clasp shut on unsuspecting flies. If you don't often have insects passing through your home, you'll need to feed your fly trap yourself with dried blood worms (which you can purchase at your local pet shop). Be sure to stimulate the trigger hairs inside the trap to ensure the plant digests them as though it has made a live catch. Gross? Perhaps. Satisfying? Absolutely.

- **Watering:** While tap water may sustain them for a short time, it's not the fly trap's favourite refreshment. Leave a small pot on your windowsill or in the garden to collect rainwater, which they prefer.
- **Light:** Keep your fly trap in the sunniest part of your home, as they like to bask in direct sunlight.
- **Size:** Healthy Venus fly traps can grow up to 13 cm (5 inches) in diameter. Each trap can be up to 2.5 cm (1 inch) across.
- **Pet friendly?** Yes, so long as they don't consume the plant. Keep well out of their reach.
- **Top tip:** Venus fly traps are dormant during the winter. You'll need to mimic their natural habitat by moving them to a cooler part of your home.

Welcome
to the jungle.

A BEAUTIFUL PLANT IS LIKE HAVING A FRIEND AROUND THE HOUSE.

Beth Ditto

SPIDER PLANT
Chlorophytum comosum

The spider plant – also known as the spider ivy, ribbon plant and airplane plant – gained its "spider" moniker from the "babies" that grow along the runners from the main bush. These offshoots look like tiny, green versions of our eight-legged friends.

- **Watering:** Spider plants like to live in moist – but not saturated – soil. During winter, cut right back, as they need time to rest and dry out before the spring.
- **Light:** These laidback plants aren't fussy. While they love to wallow in direct sunlight, they'll thrive in almost any lighting conditions.
- **Size:** In hanging baskets they can grow up to a metre (3 feet) long.
- **Pet friendly?** Spider plants aren't toxic, but can cause discomfort to pets if they are ingested.
- **Top tip:** Without attentive pruning, spider plants can quickly spiral into a miniature labyrinth. Trim regularly to keep them neat and tidy.

KENTIA PALM
Howea forsteriana

Kentia palms have a rather grand history – not only were they used to decorate first-class accommodation aboard the *Titanic*, but they were also a favourite of Queen Victoria. Thankfully, all this adulation hasn't gone to their heads – so long as they have a little light and hydration, a kentia will be just as happy in your lounge as it was in royal quarters. Treat your palm with a little fertilizer to prevent any yellowing leaves and this plant will soon become the crowning glory of whichever place it calls home.

- **Watering:** Kentia palms are prone to root rot, so make sure they have adequate drainage. Water when the top 5 cm (2 inches) of soil becomes dry, and less often in the winter.
- **Light:** These palms grow well in low light conditions, so they're a great way to liven up a darker corner of your home.
- **Size:** Kentia palms can grow up to a whopping 3 m (10 feet), so are best suited to homes with plenty of space for them to flourish.
- **Pet friendly?** Yes, they're non-toxic.
- **Top tip:** Regular misting will ensure your palm has the humidity it needs to keep its fronds fresh.

YOU CAN'T
BUY HAPPINESS,
BUT YOU CAN BUY
PLANTS, AND THAT'S
PRETTY MUCH THE
SAME THING.

WHENEVER I FIND A PLANT
THAT IS BEAUTIFUL AND
ODD, I BUY IT.

MARTHA STEWART

CHEESE PLANT
Monstera deliciosa

The lustrous leaves of the monstera are filled with distinctive holes, which have earned it the nickname: "Swiss cheese plant". These large holes serve a very practical purpose in the tropical rainforests (from which they originate), by allowing the plant to withstand heavy rain and winds. Wild monstera plants grow a delicious fruit, which apparently tastes like a combination of strawberries, mango, passion fruit and pineapple – yum! Unfortunately, your home-version is unlikely to bear fruit, but give it the love and attention it deserves and it *will* bear leaves green and glossy enough to make any plant parent proud.

- **Watering:** Allow the top 5 cm (2 inches) of soil to become dry before watering.
- **Light:** A light room with plenty of shade is the cheese plant's perfect home.
- **Size:** Monstera plants can grow to over 3 m (10 feet). To encourage their growth, push the brown aerial roots down into the soil.
- **Pet friendly?** Yes, it's non-toxic.
- **Top tip:** If your cheese plant is looking more cheddar than Swiss, encourage its leaves to develop more holes by moving it to a lighter environment. Alternatively, offer the plant a little fertilizer.

I'M SORRY FOR WHAT I SAID WHEN I WANTED PLANTS.

I'M ALWAYS TOUCHING PLANTS AND VIBING WITH THEM.

GZA

CHINESE MONEY PLANT
Pilea peperomioides

The Chinese money plant – so called for its coin-shaped leaves and purported ability to attract wealth and good fortune – makes for an extremely agreeable housemate. Somewhere between juicy succulent and leafy sensation, these cheery-looking plants are native to Southern China and became popular in Britain in the late twentieth century. Money plants are so easy to propagate that they're also known as the "pass it on plant"! New growths from your Chinese money plant will make wonderful gifts for friends and family.

- **Watering:** Mist your plant regularly to encourage photosynthesis, but beware overwatering – money plants are prone to root rot.
- **Light:** These plants are extremely sensitive to the sun. You'll find you need to rotate your pot a few times a week to avoid all the leaves growing in one direction. They like light rooms, but not direct sunshine.
- **Size:** In the space of five–ten years, a money plant can grow up to half-a-metre high (20 inches) – and just as wide!
- **Pet friendly?** Yes.
- **Top tip:** If you accidentally sever a stem, don't despair! Place it in compost and you can easily grow a brand-new baby plant.

REASONS I LOVE MY PLANT:

1. IT'S A PLANT.

HOW STRANGE THAT NATURE DOES NOT KNOCK, AND YET DOES NOT INTRUDE!

EMILY DICKINSON

CORN PLANT
Dracaena fragrans

In ancient Africa, corn plants were a symbol of good fortune – as far back as 1000 BC, northern Tanzanian tribes used them to mark their holy places. These beauties grow from one or two thick stems and fan out into long, narrow leaves, giving them the appearance of a miniature palm tree. Pour out a cocktail, pop on some unnecessary indoor sunnies and enjoy the feel of a tropical escape in your own front room.

- **Watering:** These fancy fronds love to be misted but hate too much water. Give them a liberal spritz each day, but limit watering to once a week.
- **Light:** Corn plants grow quickly in bright light, but become stronger in shade; keep your plant somewhere it will receive both so that it can fully flourish.
- **Size:** They can grow up to a statuesque 15 m (50 feet), but most house versions will usually mature at a happy 1.5 m (5 feet).
- **Pet friendly?** Corn plants are mildly toxic if ingested, so keep pets at bay.
- **Top tip:** These strong growers need room to thrive. Repot every couple of years to see them reach for the sky.

#SERIALPLANTER

WHOEVER LOVES AND
UNDERSTANDS A GARDEN WILL
FIND CONTENTMENT WITHIN.

Chinese proverb

PARLOUR PALM
Chamaedorea elegans

It can take decades for parlour palms to stretch up to their full height, so if you're looking to cultivate a jungle in a jiffy, this isn't the plant for you. If you have patience, however, and enjoy investing in a plant long-term, there's much to love about the Victorians' favourite houseplant. They're elegant, easy to care for, and can be with you for many years. Fronds from a parlour palm can last for up to 40 days after being cut away from the main plant, so they're a great choice if you enjoy creating wreaths or exotic flower arrangements.

- **Watering:** Give your palm plenty of drainage and water only when a finger pressed into the soil comes out dry. In winter, they need very little hydration.
- **Light:** Parlour palms aren't fussy. They can grow extremely well in a shaded room. Direct light may scorch their leaves.
- **Size:** Though it's a slow grower, the parlour palm can creep up to a height of 1.2 m (4 feet) if cared for well.
- **Pet friendly?** Yes, this plant is non-toxic.
- **Top tip:** Parlour palms hate to be disturbed! Repot them only when absolutely necessary.

GREEN IS THE NEW BLACK.

WE CAN'T ESCAPE
OUR ANCIENT
HUNGER TO LIVE
CLOSE TO NATURE.

DIANE ACKERMAN

PEACE LILY
Spathiphyllum wallisii

The beautiful white bloom of this tropical plant earned its name from its resemblance to white flags of peace and, indeed, it's a rather peaceful plant to care for. Peace lilies are great at removing nasty pollutants from the air, improving the air quality of your home. You might be surprised to learn, however, that peace lilies aren't lilies at all – they're actually related to their fellow houseplants, philodendrons and anthuriums.

- **Watering:** Water your peace lily with some restraint, aiming to sprinkle hydration over your soil, not to saturate it. They love a good misting, so spritz every few days.
- **Light:** A combination of both light and shade will keep your peace lily the happiest.
- **Size:** Peace lilies typically grow up to 1 m (3 feet) tall.
- **Pet friendly?** No, they're toxic to pets; keep out of their reach.
- **Top tip:** If your peace lily fails to produce its signature bloom, it's probably not getting enough light. Move it somewhere slightly sunnier and reap the rewards of a new bud.

GROW
GOOD
THINGS.

IF ONE TRULY LOVES NATURE, ONE FINDS BEAUTY EVERYWHERE.

Vincent van Gogh

BOSTON FERN

Nephrolepis exaltata "Bostoniensis"

Ferns are needy plants. They're extremely easy to kill, whether through too little water, too much water, being too cold, or excess warmth, so are among the trickier varieties of houseplants to care for. Boston ferns, however, aren't like their sensitive brothers and sisters and can often be revived from the brink of death with just a little bit of TLC. Plant them in a hanging basket and enjoy the drapery of their verdant, green fronds.

- **Watering:** Aim to keep the soil of your Boston fern moist at all times.
- **Light:** A shady spot in a sunny room will be the happiest home for your fern.
- **Size:** Treat it well and your Boston fern could grow up to 1.8 m (6 feet) wide.
- **Pet friendly?** Yes! Your pet and your plant can live in happy harmony.
- **Top tip:** These ferns lap up humidity; place them in a steamy bathroom and treat them to a daily spritzing for the best results.

#GREENTHUMBCREW

EVERY TIME I
LOOK AT MY
HOUSEPLANT
IT BRINGS ME
SO MUCH JOY.

ZOELLA

JADE PLANT
Crassula ovata

The jade plant is commonly known by its nickname – "the money plant" – as for centuries people have spoken of its ability to attract wealth into the life of its owners. And luck-seeking would-be owners are spoiled for choice, as there are over 1,400 varieties of jade plant, each with a more intriguing name than the last: the lucky plant, dwarf jade, true jade, Gollum jade... are you sold yet?

- **Watering:** Jade plants enjoy a light watering when their soil is fully dry, and even less in the wintertime.
- **Light:** A spot that offers around 5 hours of indirect sunlight is perfect.
- **Size:** When pruned, jade plants can reach a height of 1 m (3 feet); left to their own devices, some specimens can tower at 1.8 m (6 feet).
- **Pet friendly?** While jade plants aren't toxic, they can cause nausea, so keep away from your pets.
- **Top tip:** Jade plants are very sensitive to repotting – you should only do so when they have completely outgrown their home.

I'm a plant parent. What's your superpower?

I HAVE A LOT OF HOUSEPLANTS, WHICH ARE KIND OF LIKE MY CHILDREN.

BEL POWLEY

ALOE VERA
Aloe barbadensis

Aloe vera is a charitable plant; it can be used to treat skin conditions, heal wounds, relieve heartburn and even be whipped up in smoothies – the benefits of having this spiny friend in your home are endless! Luckily, they're very hardy plants too and will proudly withstand a fair amount of neglect, making them a great choice for first-time plant parents or those with particularly busy lifestyles.

- **Watering:** Aloe is a succulent so be careful not to overwater. A little hydration when the top 2.5 cm (1 inch) of soil is dry will suit them perfectly.

- **Light:** Bright but indirect sunlight is this juicy plant's preference.

- **Size:** You may be surprised to learn that a single aloe spike could grow as tall as 90 cm (35 inches) during your plant's lifetime and, if encouraged, the whole plant can grow to a metre (3 feet) across.

- **Pet friendly?** Aloe is toxic when ingested by pets, so keep out of their reach.

- **Top tip:** The thick spines of the aloe plant make it rather top heavy. To help it stay balanced, be sure to use a heavy pot – concrete planters are a particularly good choice.

I GET BY WITH A LITTLE
HELP FROM MY PLANTS.

NATURE DOES NOT
HURRY, YET EVERYTHING
IS ACCOMPLISHED.

LAO TZU

FIDDLE LEAF FIG
Ficus lyrata

The lyrical name of this plant comes from the shape of its leaves. However, despite its jaunty namesake, the fiddle leaf fig is considered a sinister plant in its native West African rainforests. Frequently beginning life on the branches of another plant, the fiddle leaf needs to reach its roots down to the ground in order to mature. To do this, it envelops and strangles the host plant, and then feasts upon any nutrients left in the soil. Thankfully, they're much more peaceful when grown in the home, and its large shady leaves are really quite serene.

- **Watering:** Overwatering can cause a lot of damage and make your plant drop its leaves. Proceed with caution, and always ensure soil is dry before watering.

- **Light:** Fiddle leaf figs love short blasts of afternoon sun, but prefer indirect sunlight for most of the day.

- **Size:** Depending on the care they receive, these plants can grow up to 3 m (10 feet) tall.

- **Pet friendly?** No, these trees are toxic. Keep them away from pets.

- **Top tip:** Once your plant reaches your ideal height, prune them from the top to discourage further growth.

EAT,
SLEEP,
PLANT,
REPEAT.

ADOPT THE PACE
OF NATURE: HER
SECRET IS
PATIENCE.

RALPH WALDO
EMERSON

RUBBER PLANT
Ficus elastica

With their deep-green shiny leaves, a well-cared-for rubber plant can look so perfect, you might double-take before you realize it's not fake. Don't be fooled by their picture-perfect appearance though – these plants are seriously tough. In their native habitat of South Asia, they can grow up to 61 m (200 feet) tall, and in India they have been cultivated to form living bridges thanks to the strength of their roots. At home, they're a humbler plant, and keeping them in a small pot will ensure they stay a manageable size, unlike their enormous international siblings.

- **Watering:** Water your rubber plant once its soil becomes slightly dry to the touch.
- **Light:** A pleasantly sunny spot will suit this plant very well.
- **Size:** The size of your rubber plant depends on the pot you keep them in, but generally houseplant versions are sold at either 30 or 60 cm (12 or 24 inches).
- **Pet friendly?** No, this plant is toxic if ingested.
- **Top tip:** If your plant starts shedding its leaves, you're watering it too much. Move it somewhere sunnier and hold off on the hydration!

I'VE NEVER MET A PLANT I DIDN'T LIKE.

I FELL IN LOVE WITH
FLORA OF ALL TYPES,
ESPECIALLY FERNS. LOVED
THE SPARSE STRUCTURE
AND REPETITION OF SHAPE.

Jack Dorsey

WEEPING FIG
Ficus benjamina

The weeping fig lives up to its name – this tree-like plant is a sensitive soul. Move your plant around the home too much and they will go into shock; water them too much and they'll die; don't water them enough and their leaves will turn crispy. But, if you're willing to tread gently, these plants will be beautiful additions to your home, with graceful arching branches and slender, pointed leaves.

- **Watering:** Water when the top of the soil is dry to the touch. Lower soil should be kept moist, but at no point should your weeping fig be sitting in water.
- **Light:** The darker the leaves of your fig, the less light it can endure. Keep your plant in a space where it will receive more shade than sunshine.
- **Size:** Miniature types will grow to 1 m (3 feet); regular plants can reach 3 m (10 feet).
- **Pet friendly?** No, this plant is mildly toxic if ingested.
- **Top tip:** Invest in a mister to keep your fig happy. These plants absorb moisture through their leaves, so they love humid air to keep their fronds fresh.

The plant life
chose me.

AS SOON AS I STARTED
COLLECTING HOUSEPLANTS,
I COULDN'T STOP. HONESTLY,
I CAN NEVER HAVE ENOUGH.

CONNOR FRANTA

DEVIL'S IVY
Epipremnum aureum

Don't be put off by this creeping plant's hellish name –
it's really a heavenly house guest. Hang it somewhere
high, and your ivy's tendrils will grow long and leafy.
If you want to create a truly spectacular houseplant
display, experiment with training it up a wall to create
your very own home jungle.

- **Watering:** This creeper is tough, so don't be nervous
  about underwatering; you're far more likely to cause
  damage by over watering. Only water when the top
  0.5 cm (0.2 inches) of soil is dry.
- **Light:** Devil's ivy earned its name from her love of
  the darkness, so it will be perfectly happy in a shady
  spot in your home.
- **Size:** This plant's leafy tendrils can grow up to 1.8 m
  (6 feet) long.
- **Pet friendly?** No, this ivy really *is* devilish to cats and
  dogs, and is toxic if ingested.
- **Top tip:** This communicative plant will tell you when
  it's ready to be re-potted – just look out for aerial
  roots reaching out from the soil or breaking through
  the pot.

QUEEN OF THE
URBAN JUNGLE.

PLANT EXTRA, BECAUSE ONCE YOU START YOU WILL ONLY WISH FOR MORE.

Jennifer Garner

ZZ PLANT
Zamioculcas zamiifolia

ZZ... This plant is so easy to look after, you could practically do it in your sleep! Unfussy when it comes to light, thirst and humidity, this houseplant will grow with only the most basic care. ZZ plants are known for their superior air purifying skills, and their shiny leaves, which brighten up any room they feature in; what did we do to deserve them?!

- **Watering:** Only water your ZZ when the top 2.5 cm (1 inch) of soil is dry.
- **Light:** ZZs love bright light but will grow fine without it, so pop them wherever looks best.
- **Size:** Your ZZ's stems will reach a maximum height of 30 cm (12 inches).
- **Pet friendly?** No, these plants are poisonous to cats and dogs.
- **Top tip:** If your plant is looking a little brown, it could be a sign of dry air. Give them a regular light misting and they'll soon be back in the green.

#GREENAF

I'VE YET TO FIND A DESIGN
CONUNDRUM THAT CAN'T
BE SOLVED WITH PLANTS.

JUSTINA BLAKENEY

SUCCULENTS

Succulents are among the most cheerful of houseplants, with their jazzy array of textures, shapes, colours and sizes offering a wealth of possibilities to your interior décor. Because succulents store moisture in their leaves, they're a great choice for absent-minded waterers, and their small stature means there's space for them in any home.

- **Watering:** Only water your succulents when their soil is dry and stop completely during the winter – they will create a reservoir inside their colourful bodies to use in the cooler months, when they become dormant and rely on their summer water supply.

- **Light:** Succulents thrive in strong, bright light. If your home is on the darker side, search for pigmented plants; the greener the succulent, the hardier they are in the shade!

- **Size:** Generally speaking, succulents are small plants, and look wonderful displayed in groups.

- **Pet friendly?** Many succulents are mildly toxic to cats and dogs. To be safe, make sure you research specific varieties for tailored advice.

- **Top tip:** Soil is everything for succulents – be sure to choose a mixture designed specifically for them, which includes sand, perlite or pumice to keep them feeling at home.

POT IT LIKE IT'S HOT.

NATURE IS NOT A PLACE TO VISIT. IT IS HOME.

Gary Snyder

CACTI

Is there any houseplant more iconic than the cactus? Often collected for their funky shapes as well as for their greenery and flowers, a well-kept cactus can be with you for decades. Seeking inspiration? Look to the gardens of Morocco's magnificent Jardin Majorelle for prickly paradise.

- **Watering:** Overwatering is just about the only way to kill a cactus, so ensure you only water your plant when the soil is completely dry – and remember, too little is always better than too much.

- **Light:** As they're accustomed to growing in deserts, cacti need lots of sunlight. Beware direct sun, however, as this can burn your plant.

- **Size:** Cacti fluctuate greatly in height, with different varieties growing between 2.5 cm (1 inch) and 19 m (63 feet) high. The Mexican giant cardon is the world's largest variety. Growth is slow in cacti, however, and some varieties can take as long as ten years to grow just 2.5 cm (1 inch).

- **Pet friendly?** Pets – like humans – don't really like being pricked. Keep them at a safe distance.

- **Top tip:** Cacti need a yearly rest. Don't water them over the winter months, when their inner store provides them with plenty of hydration.

PLANT
MORE SEEDS.

COLLECT THEM, FEED
THEM, CLUSTER THEM,
LOVE THEM. PLANTS
WILL LOVE YOU BACK.

Genevieve Gorder

RADIATOR PLANT
Peperomia

Radiator plants have many aliases, all of which are unaccountably adorable – pin cushion plants, the happy bean, the baby rubber plant – and their many varieties all make for cute additions to your home. The *argyreia*, which has leaves resembling the rind of a watermelon, is a particular favourite. Interestingly, and as their botanical name suggests, *Peperomia* belong to the same plant family that produces peppercorns.

- **Watering:** Radiator plants enjoy a thorough watering any time their soil becomes dry – usually once a week.
- **Light:** Keep things bright for your radiator plant. They can endure varying levels of sunlight, but sitting in the dark will stop them from growing.
- **Size:** These pocket-sized plants rarely grow taller than 20 cm (8 inches).
- **Pet friendly?** Yes! Your pooch and your *Peperomia* can coexist in harmony.
- **Top tip:** Not sure where to place your plant? Radiator plants thrive in high humidity, making your bathroom the perfect home.

Let it grow.

A LITTLE BIT OF GREEN
HAS A GREAT EFFECT
ON HAPPINESS.

BOBBY BERK

CAST IRON PLANT
Aspidistra elatior

As their name suggests, this plant is the ultimate fighter – you'd have to be extremely gifted in the art of plant neglect to kill these babies off. Native to Japan, these tall and slender beauties have large, paddle-shaped leaves and dainty long stems that look splendid in unfrequented corners of the home. Thank them for their hardiness by occasionally cleaning their leaves with a soft cloth (see p.14) and you'll be friends for life.

- **Watering:** Leave your plant alone. Cast iron plants require sparse watering and can survive for long spells with no drink at all.
- **Light:** Indirect but bright light is the iron plant's preferred setting.
- **Size:** These plants can mature up to 1 m (3 feet) tall.
- **Pet friendly?** Yes, the cast iron plant is tough but non-toxic.
- **Top tip:** Cast iron plants are slow growers (too busy being tough guys), but feed them a little fertilizer once a month during the spring and summer and you might just speed up their progress.

MAJESTY PALM
Ravenea rivularis

All hail the majesty palm! This splendid green goddess is a variety of palm tree, native to South Central Madagascar. These palms aren't as common as other varieties, but for tropical plant enthusiasts they make a wonderful choice.

- **Watering:** Majesty palms like their soil moist. During the summer they may need watering as much as twice a day. Ensure you have adequate drainage to prevent root rot.
- **Light:** Majesty palms are all about the sunshine. To grow to their full potential, they'll need between 6 to 8 hours of bright light each day.
- **Size:** Wild majesty palms can tower up to 24 m (80 feet), but don't panic – indoors they're extremely slow growers, so they won't punch a hole in your ceiling any time soon.
- **Pet friendly?** Yes, majesty palms are non-toxic.
- **Top tip:** Don't be shy: fertilize! Majesty palms love some extra TLC, and regular fertilizer will prevent potential nutrient deficiency.

PLANTS BRING SUNSHINE
TO MY SOUL.

HAVE NOTHING IN YOUR
HOUSE THAT YOU DO NOT
KNOW TO BE USEFUL, OR
BELIEVE TO BE BEAUTIFUL.

William Morris

VELVET CALATHEA
Calathea rufibarba

Treat yourself to a *Calathea rufibarba* and you might find you spend all day rubbing its velvety leaves, which are a stripy, crinkled green on one side and a plush purple underneath. Also known as the jungle calathea, these plants symbolize fresh beginnings; in fact, the expression "turn over a new leaf" comes from this plant's curious tendency to roll its leaves up at night and unfurl them again at dawn.

- **Watering:** *Calathea* enjoy moist soil, but if they sit in water for too long, they'll die. Water them little and often.
- **Light:** The brighter the indirect light your calathea soaks up, the finer her foliage will be.
- **Size:** A mature *Calathea rufibarba* may grow up to a metre (3 feet) tall.
- **Pet friendly?** These plants are non-toxic, so your kitty and your calathea will soon be firm friends.
- **Top tip:** These tropical plants don't like the cold. Try to keep them in an environment of around 18–23°C (64–73°F), though they can cope with 15°C (59°F) in the winter.

SORRY, I CAN'T COME – MY PLANTS NEED ME.

GREEN FINGERS ARE THE EXTENSION OF A VERDANT HEART.

Russell Page

ASPARAGUS FERN
Asparagus setaceus

There's something whimsical about the soft, feathery leaves of the asparagus fern. Light and fluffy in appearance, these plants are bushy in childhood but soon slim down as they mature, developing vines and more slender stems. Strangely enough, these plants aren't ferns at all – they are, however, related to the edible asparagus. Plant them in a hanging basket to really highlight their fairy-like looks.

- **Watering:** Don't let your fern's soil dry out – these guys like to drink! Regular misting will help to keep them hydrated.
- **Light:** As they naturally grow under the shade of a tropical canopy, asparagus ferns are happy in partial shade or indirect light.
- **Size:** Mature ferns may grow up to 0.6 m (2 feet), but this happens relatively slowly.
- **Pet friendly?** No. Despite its edible namesake, the asparagus fern is toxic to pets.
- **Top tip:** As soft and fluffy as these plants look, please don't touch them! Too much handling will turn their greenery brown.

PROUD
PLANT
PARENT.

JUST LIVING IS NOT
ENOUGH. ONE MUST
HAVE SUNSHINE,
FREEDOM AND A
LITTLE FLOWER.

HANS CHRISTIAN ANDERSEN

SWEETHEART PLANT
Philodendron scandens

The sweet, heart-shaped leaves of the philodendron look beautiful cascading from a hanging basket. Treated well, the vines of this climbing plant can grow to up to 3 m (10 feet) long, with each leaf up to 25 cm (10 inches), making it a real statement for any foliage-filled home.

- **Watering:** The soil of your sweetheart plant should be almost entirely dry between watering, except in winter, when you should allow it to dry out completely. Weekly watering suits them well.

- **Light:** Sweetheart plants can flourish in most kinds of light, but keep them away from direct sun, which will scorch their leaves.

- **Size:** These plants grow remarkably fast in the right conditions, so although young plants tend to be around 15 cm (6 inches) tall, you could soon find yourself with a bountiful 25 cm (10 inch) trail of leaves.

- **Pet friendly?** No. The sap of this plant can cause skin irritation or burn the mouth of your pet if ingested.

- **Top tip:** If your sweetheart plant becomes leggy, it's a sign that it's not getting enough light. Move it to a sunnier spot and watch it thrive.

LAST WORD

Not only do plants transform a home, but they can transform a person too. Taking the time to nurture something, care for it, and watch it grow, can be a calming and restorative experience. Forging a relationship with the plants around your home is a signal to yourself and the world that you are investing in your environment, and in the physical and mental benefits that plants can offer you.

Hopefully the tips in this book will serve you well, and help you find your very own green gang. Whether that means a fortuitous relationship with your Chinese money plant, coexisting with your proud parlour palm or getting zen with a peace lily, finding the plant that best fits you is a step toward finding yourself. So, armed with our tips and tricks, go forth, gather and nurture a plant family of your very own.

ALL YOU NEED IS LOVE.

AND A PLANT.

PLANT INDEX

IMAGE CREDITS

If you're interested in finding out more about our books, find us on Facebook at **Summersdale Publishers** and follow us on Twitter at **@Summersdale**.

www.summersdale.com

THE GREATEST SHOW ON EARTH

The first of the modern Olympic Games, took place in Athens in 1896. Among the events contested was the marathon. This was run over the route from Marathon to Athens and was won by the Greek Spiridion Louis. The inspiration for this classical race was the wish to recall the great feat of Pheidippides who, in the 490 B.C., brought the news to Athens of the incredible victory won by Miltiades over the Persians.

THE vast stadium in Berlin, built to foster sport and friendship, turned suddenly into a cauldron of hatred. Before the eyes of the world, dictator Adolf Hitler was using the Olympic Games to further his gloating boasts of German superiority.

Everywhere, German spectators raised their arms in the Hitler salute. Military bands played Hitler's music. German athletes were greeted with fervent shouts of "Heil Hitler," and members of the Hitler Youth, young converts to the Nazi creed, strutted arrogantly about the arena.

Hitler's vain boasts were shattered by one of the negroes he despised. Sprinter Jesse Owens, of the United States, won four gold medals and set up three Olympic records to demonstrate to the world that Hitler's ideas were nonsense.

But for all his lightning speed on the track. Owens will be remembered in history for one superb gesture of sportsmanship with which he restored faith in the Olympic ideal of fostering friendship.

In the long jump Owen's chief rival was a German, Lutz Long, whom the Nazis idolized and expected to

A

win. During the contest Long developed cramp, and thought he would have to retire. Owens hurried to his side, and massaged his legs until he was fit to jump again.

Long made his third and final jump—and the Germans screamed with delight as he took the lead. But when Owens jumped again he not only won the event but set up a record that stood unbeaten for twenty-four years.

The Olympic flame is kindled in Greece because the games had their origin in the village of Olympia nearly three thousand years ago.

The valley of Olympia was held sacred to Zeus, king of the gods. Here was his shrine and here was set up his superb statue, carved by Pheidias in gold and ivory, which was among the Seven Wonders of the World. At the altar of Zeus sacrifices were made, and a sacred fire burned there perpetually.

At four-year intervals Zeus kept high festival. Wars were suspended, and fighting men made a pilgrimage to Olympia. From these gatherings of warriors the Olympic Games were born.

For five days, a thrilling festival of sport was held. Chariots, with horses teamed four abreast, charged through the arena, turning hairpin bends at each end in a succession of thrilling contests. There were foot races, wrestling and boxing matches, and fierce duels to test an athlete's all-round skill.

In one event—the pankration, a form of all-in wrestling—a contestant was not considered beaten until he acknowledged defeat. One wrestler died at the moment of triumph—just as his opponent surrendered.

It is impossible to exaggerate the importance of the Olympic Games in the life of the Greeks. They calculated their dates by the Olympiad, the period of

One of the most spectacular races of the ancient Olympic Games was the race of the four horse chariots. It was a highly exciting race which fired the spectators with enthusiasm and offered an unforgettable spectacle of strength, courage and ability.

Right: The Olympiad held in London in 1908 had its historic moment. During the marathon the Italian Dorando Pietri, after having dominated the whole race, fell, exhausted by his superhuman effort, only a little way from the finishing line. He was disqualified because he was helped along by some of the officials, but he was acclaimed as the moral victor of that extremely exhausting contest.

four years from one festival to the next. An Olympic victory was the greatest triumph any mortal could achieve. Victors received only one prize—a crown of olive branches cut from a sacred grove by a boy with a golden knife. When they returned home they were fêted like gods. Sometimes a breach would be made in a city wall so that a victor might drive home in triumph and be received as a conqueror.

The Games survived until A.D. 394 when, during the 293rd Olympiad, they were abolished by the Christian Emperor Theodosius as a heathen observance.

For 1,600 years the Olympic spirit lay dormant. Then, at Athens on April 6, 1896, the first of the modern Games were declared open. They were the inspiration of a French nobleman, Baron Pierre de Coubertin who saw in them a means of fostering peace.

He summed up the Olympic ideal in a few historic words: "The important thing in the Olympic Games is not winning but taking part. The essential thing in life is not conquering but fighting well."

Since then, with two interruptions for war, the Olympic Games have been held every four years. growing in size and influence with every celebration,

Nowadays they are held in a different city each time, and include not only athletic events but almost every international sport for men and women.

Anybody nominated to represent a country may take part. There are no barriers of class or age. The only—but vital—rule is that competitors must not earn their living at or receive money from sport. There are no prizes, but the Olympic gold medal, presented to the winner of each event, is the most coveted award any sportsman can hope to gain. A silver medal is given to each runner-up and a bronze medal to the third.

Many times athletes such as Jesse Owens have shown the truth of de Coubertin's message—that to take part is more important than to win.

At the Los Angeles Olympics of 1932, the bronze medal for third place in the 110 metres hurdles was given to Jim Keller, of the United States. A newsreel film later showed that Britain's Donald Finlay was third. Keller insisted on handing Finlay the medal.

In the 3,000 metres steeplechase at the same Olympics, officials made an appalling blunder. Competitors were allowed to run one lap of the track too many. Had the race ended at the correct distance, the second and the third places might have been reversed. A British runner, Evenson, was awarded the silver medal, but wanted to hand it to the third man, an American named McCluskey.

McCluskey refused to listen. "Who can say what might have happened?" he asked. "I was third over the finish—and a race has only one finish."

If the Olympic Games promote sportsmanship, they also produce some fantastic feats of courage. Karoly Takacs, of Hungary, was the world pistol shooting champion in 1939. During the war he lost his shooting arm. For most sportsmen that would have been the end, but Takacs refused to give up. He learned to shoot with his left hand—and in 1948 and 1952 won the Olympic titles against the finest marksmen in the world.

And who can deny the courage of New Zealand's Murray Halberg, crippled by polio as a child until he was left with a withered left shoulder and arm? Halberg determined to become a top-class athlete, and at the 1960 Games in Rome he outstripped the world's best to win the 5,000 metres.

At the ancient Olympic Games women were banned from competing—and even barred from watching, on pain of death. Today the performances of women competitors excite as much interest as those of the men. Certainly at recent Olympics they have often stolen the show.

Modern Olympic games have had their moments of drama and excitement as when a little Italian pastrycook from the Isle of Capri staggered wearily along the road towards London's White City stadium. For nearly twenty-six miles, in his red shorts and white vest, he had run his heart out in one of the most heroic races the world has known.

Now, as he stumbled through the gateway and on to the running track for the final few hundred yards, thousands in the packed stadium roared encouragement. But the little Italian could hardly hear them. He was dazed, near to collapse.

He should have turned left towards the finishing line. Instead he turned right, bewildered. Officials and policemen guided him back. Then Dorando Pietri, the little pastrycook, fell to the ground, his body drained of strength.

Frantic officials, excited by his courage, helped him to his feet and Dorando, scarcely knowing where he was, dragged his way to the finish. Following him, 150 yards

behind, came an American runner, John Hayes.

At first Dorando Pietri was fêted as the victor. Then the Americans protested that he was helped to finish, and the Olympic Games marathon race of 1908 was awarded to Hayes.

For weeks argument raged over whether Dorando— he came to be known by his Christian name—might have won the race without help. The day after his gallant performance, Queen Alexandra, to tumultuous cheering, presented him with a gold cup as consolation for his defeat.

Few people now remember the winner of that epic marathon. But the name of Dorando goes down in history as the immortal loser, a man whose courage won him greater glory than victory itself.

No race in the Olympic Games creates more excitement than the marathon. It takes its name from the plains of Marathon, in Greece, where the Athenians won a famous battle in 490 B.C.

After that battle, according to legend, an Athenian herald named Pheidippides ran the twenty-five miles to Athens with news of the victory. He fell dead uttering, in Greek, the words: "We have conquered." The poet Browning commemorated this exploit in one of his poems.

When the Olympic Games were revived in 1896, the marathon was included. Its distance—26 miles 385 yards, the only Olympic distance not based on metres—was not fixed until 1908. Many people believe, quite wrongly,

that this unusual distance is that from Marathon to Athens. In fact, the 1908 race started at Windsor Castle and should officially have been "about twenty-five miles." When the course was measured it was found to be exactly twenty-six miles to the stadium entrance.

It was then decided that the ideal finishing point was in front of the royal box—385 yards farther on. Since then, 26 miles 385 yards has become the accepted marathon distance.

The marathon has always been a race of drama. Runners have died in it. Others have so taxed their strength that they have never run seriously again. It is a race that topples champions, and makes unknown runners into world heroes.

Twice in recent years the Olympic marathon has been won by men who have never raced in a marathon before. In 1952 at Helsinki, Emil Zatopek, one of the greatest athletes the world has known, completed an incredible hat-trick of victories by winning the 5,000 metres, 10,000 metres and the marathon.

In all three races the Czech army major set up Olympic records—and after the marathon demonstrated his superb physical fitness by calmly eating an apple and chatting to friends while he waited for the others to finish.

At every Olympic Games, the marathon gold medal becomes harder to win—and it is always full of surprises and excitement for the spectators.

Originally the Olympic Games only consisted of a few races, of which the majority were in the field of light athletics. In the modern Olympic Games, on the other hand, the various contests cover the whole field of sport: athletics, water sports, basket-ball, fencing, equestrian sports and many others including winter sports.

4

THE SHIP THAT NEVER SAILED

Above: An artist's impression of the *Vasa*, as she would have looked.
Below: The great ship as she looked after being brought to the surface in 1961.

O N a fine Sunday afternoon in August, 1628, a huge and jolly crowd lined the waterfront of Stockholm's harbour. They had come to see the climax of an event which had been the talk of the town for months past—the launching of the *Vasa*, Sweden's mightiest warship.

Named for the royal family, she was as fine a galleon of 1,400 tons as ever would sail the seas. She was 210 feet overall with 64 gleaming cannon, a wonderfully ornamented hull, and you had only to look at her to believe that her country would be invincible while she rode the seas.

The breeze filled the *Vasa*'s canvas and

Above: *Vasa's* anchor on view outside the warship's last resting place.
Below: Inside the building which houses the *Vasa*, water jets play
ceaselessly on the old ship's oak timbers.

she slipped into the harbour to a great roar of approval from the watching crowd. The breeze changed as she sailed out into the centre of the water. It was not much of a breeze by any standards, but it was a fresh breeze that was to make history.

For, inexplicably to the horrified spectators, it heeled the magnificent *Vasa* over to port and sank her like a stone in 100 ft. of water.

Small boats crowded round the instant disaster scene and picked up hundreds of survivors. But still about 50 died in the ship which, it has since been decided, was architecturally so wrong that it was doomed to disaster.

For three centuries the *Vasa* lay in her watery grave in Stockholm harbour—at first an incredible memory in the minds of seventeenth century Swedes and then, as the years passed—a long forgotten story.

Then, in 1956, a Swedish admiralty engineer, whose interest in the *Vasa* story had been stimulated by a Swedish historian, set out to pinpoint the place where the dead warship lay. After some fascinating detective work he found it, and called in the Royal Swedish Navy.

A committee was set up and after many months of study it was decided that the ship could be raised. Government, industry and interested individuals subscribed to a fund to collect the necessary three-quarters of a million pounds the project would cost, while a salvage company took up the job of directing naval divers on the sea bed.

In April, 1961, 333 years after she sank, another huge crowd gathered on the edge of Stockholm harbour to see the *Vasa* raised from the dead—a great skeleton of oak from out of the pages of history.

Experts examining the hulk found the beams and the cannon ports in near perfect condition. Even the ship's water pump was in working order. All this was possible because the Baltic Sea's high salt content does not encourage wood boring microbes to thrive, microbes which, in any other sea, would have made a disastrous meal of the *Vasa's* timbers.

On board, too, the experts found naval instruments, clothes, human skeletons, enough artifacts to stock the *Vasa* museum, which is now one of the tourist attractions of Stockholm. Close by the museum, a great building of concrete and glass was erected to house the hulk.

There, for ten years now, and for many more years to come, engines pump thousands of gallons of water every day over *Vasa's* old timbers, to prevent them from drying out too rapidly and thereby rotting away. Visitors walk around platforms inside the half-lit building while the jets spray the ship only inches from their faces.

The successful raising and subsequent housing of the great old ship is a tribute to Swedish ingenuity and resourcefulness, everywhere in evidence in the scenic city of Stockholm, Sweden's capital on the water.

NANSEN—
THE MODERN VIKING

Explorers were horrified when the Norwegian explorer told them that he planned to sail and walk over the ice-bound top of the world. How, they argued, could any man spend three winters in the Arctic and survive?

On the 26th September 1888, Fridtjof Nansen completed his great achievement of crossing Greenland. The terrible winter, which had arrived unexpectedly, prevented him from returning to his own country. He remained for many months in the little port of Godthaab, living the hard life of the Eskimos, studying their customs and learning their language.

TO the crew of the polar ship Fram it was like the coming of Domesday. Huge blocks of jagged ice were roaring down on them as they lay jammed and unable to move in the frozen wastes near the North Pole. It seemed as though the people who had called them "crazy madmen" were right.

Rushing up on deck, the Fram's captain, Norwegian explorer Fridtjof Nansen, told his men to get ready to abandon ship.

"Masses of snow and ice rushed on us, high about the rail amidships and over the deck tent," Nansen wrote in his diary later. "I saw that the tent-roof was bent down under the weight of the masses of ice. I was afraid the ice might rush over the deck, rush down the ladder, and thus imprison us like mice in a trap."

Frantically, Nansen told the crew to move all the equipment they could—dogs, sledges, kayaks and tents—

on to the steady ice on the starboard side. The Fram was now listing nearly seven degrees, and the threatening ice loomed six feet above the rail.

"The saloons and the berths were soon cleared of bags and the deck as well, and we started taking them along the ice. The ice roared and crashed against the ship's side, so that we could hardly hear ourselves speak.

"It was a fearful hurly-burly in the darkness . . . as if Domesday had come."

For hours the crew stood ready to take to the ice and try to find their way south to safety. But slowly the danger passed. The Fram was a sturdy ship, built specially for such conditions. She resisted the pressure, and found herself gradually being freed as the ice splintered all around.

Nansen saw this as his opportunity to plunge on ahead to the North Pole. He would let the Fram drift a little

Carried by the drifting ice Nansen's ship the "Fram", did not manage to reach the North Pole. The great Norwegian then disembarked on to the ice, prepared three sledges and with only one companion, his Lieutenant from the ship, started out towards his ambitious goal. But the great dream of his life was not destined to become reality.

farther forward, then he would leave her to sail back to Norway without him. So, six weeks later, he and Lieutenant Frederick Johansen, the ship's stoker, set off on the most perilous polar expedition ever undertaken.

Their departure, on March 14, 1895, came after the Fram had been drifting at sea—or at ice—for two years. Although it was not Nansen's ambition specifically to reach the North Pole, he wanted to do what no explorer had done before—to drift across the Polar seas, letting the ice carry him as far north as possible.

When Nansen had travelled to London to lay his plans before the Royal Geographical Society, they were listened to with horror. All the experts predicted disaster. Nevertheless they contributed £300 to the Norwegian's "preposterous" venture. A further £25,000 came from various scientific bodies, private donors, and the Norwegian Government.

The thought of spending possibly years away from mankind and civilization did not worry Nansen. Since childhood he had been used to being alone, and had formed the philosophy that it was only in solitary places, in isolation, that one can grow to know oneself.

Years later, speaking at a convention for schoolchildren in 1921, he summed-up his ideas on the lonely open-air and its development of character.

"It seems as though sport can no longer be carried on except in large parties," he said. "But an important part of outdoor sport should be the escape from the many, from the eternal hubbub, to the world of nature.

"The wise men of India said that every man should spend at least one hour of every day in solitude and devote it to meditation, to finding himself.

"The eternal whirlpool where men incessantly rub up against each other until they become round, smooth ciphers and men about town, is not qualified to develop individuality and character.

"Pleasures and diversions we must have, but it is important to choose sensibly those which give lasting happiness and recreation. It is in the wilderness, in the solitude of the forest, that character is formed."

It was certainly the wilderness—and a frozen one full of uncertainty and danger—that faced the two men who, as related earlier, set off overland from the Fram to travel north.

To begin with, Nansen and Johansen planned to travel forward with three dog-drawn sledges for fifty days. They reached their farthest north easily enough at the beginning of April, but it was the southbound journey to Franz Josef Land that proved really hazardous. Already some of the dogs had been killed, and the two men were aching and weary with drawing one of the sledges.

The ice was too tightly packed to give their kayaks (canoes) sailing room, and Nansen wrote: "The ice grew worse and worse. It brought me to the verge of despair. Lanes, ridges, and endless rough ice, it looks like an endless moraine of iceblocks; and this continual lifting of the sledges over every irregularity is enough to tire out giants.

"It will be slow work indeed if there be much more of this ice towards Franz Josef Land."

Nansen did not anticipate having to spend his third Arctic winter away from home, but that is what happened. Constantly beset by the ice, snow and wind, down to only three dogs, and having to burn one of their sledges for firewood, the two men made hard and painful progress.

Above: A model, made by a member of the crew, of Nansen's ship, *Fram*.

Right: Dr. Nansen, the great Norwegian explorer, in his later years.

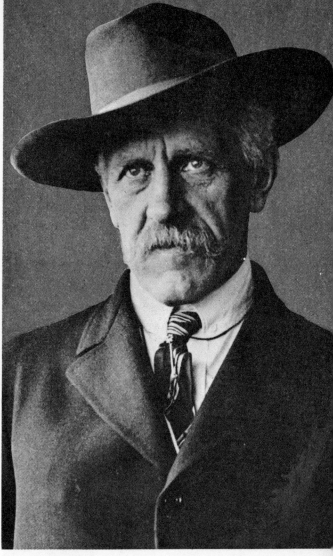

"It is getting worse and worse," recorded Nansen. "Wretched snow, uneven ice-lanes, and villainous weather stopped us. There was certainly a crust on the snow, on which the sledges ran well when they were on it; but when they broke through—and they did it constantly—they stood immovable.

"This crust, too, was bad for the dogs, poor things! They sank through it into the deep snow between the irregularities, and it was like swimming through slush for them."

Apart from all this, the explorers were attacked by polar bears and walruses, and had to guard their camp at nights against the marauding foxes which sneaked in and stole all the food and equipment they could get their teeth into.

By the following spring, however, they were able to use their kayaks, and the terrible days of sleeping all round the clock were over. Once, while spying out the land, they almost lost their little boats. The kayaks started drifting away, and Nansen had to tear off his clothes and swim off after them through the freezing water. It was, he said, "The worst moment I have ever lived through."

Finally, on June 17, 1896, just south of the Franz Josef Land archipelago, they spotted the dark form of a dog, and farther off the figure of a man. Then came a meeting as historic as the one between Livingstone and Stanley. As Nansen approached the man, the British explorer Frederick Jackson, their dialogue was charming in its simplicity:

Jackson said: "I'm immensely glad to see you."

Nansen replied: "Thank you. I also."

"Have you a ship here?"

"No, my ship is not here."

"How many are there of you?"

"I have one companion at the ice edge."

"Aren't you Nansen?"

"Yes, I am."

"By Jove! I am glad to see you!"

Nansen had done what all his critics said was impossible. He had walked into and out of the clutches of an icy death. But for all his adventures none of his childlike wonder had left him.

Nansen's biggest thrill was yet to come—his re-union with his beloved ship *Fram*. He saw her again at the port of Tromso. He was destined never to take her to polar regions again, but he wrote:

"It was strange to see again that high rigging and the hull we knew so well. When last we saw her she was half buried in the ice. Now she floated freely and proudly on the blue sea, on Norwegian waters. The meeting which followed I shall not attempt to describe."

A hero's welcome awaited him in Oslo. The guns of the old castle of Akershus thundered a salute as Nansen sailed up the fjord, and he wistfully recalled:

"That rainy morning when I last set foot on this strand. More than three years had passed, we had toiled and we had sown, and now the harvest had come. In my heart I sobbed and wept for joy and thankfulness."

Nansen was then thirty-five, and his career as an explorer was over. He had made three daring and successful voyages, now he entered upon the second and perhaps even more important stage of his life.

Rejecting the premiership of Norway on religious grounds—he refused in 1905 to join the State Church—he became his country's representative at the League of Nations in Switzerland. He was described by one journalist as being: "One of the sights of Geneva, the proudest after Mont Blanc."

When he died in 1930, Norway declared a day of national mourning for his burial. A fitting tribute for the end of a man lovingly known as "the modern Viking."

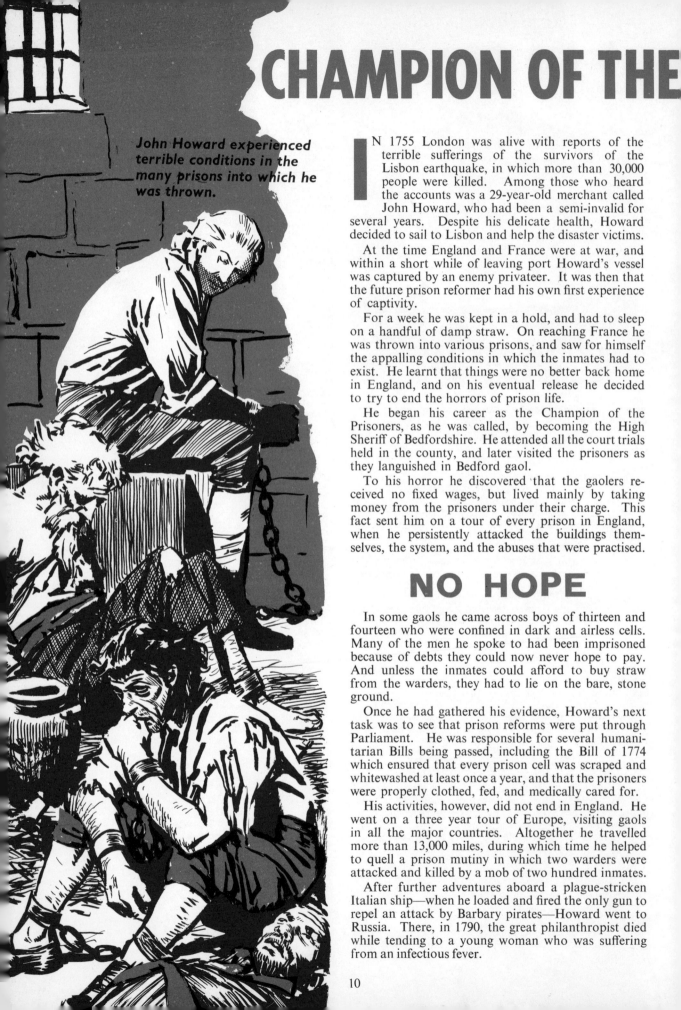

John Howard experienced terrible conditions in the many prisons into which he was thrown.

IN 1755 London was alive with reports of the terrible sufferings of the survivors of the Lisbon earthquake, in which more than 30,000 people were killed. Among those who heard the accounts was a 29-year-old merchant called John Howard, who had been a semi-invalid for several years. Despite his delicate health, Howard decided to sail to Lisbon and help the disaster victims.

At the time England and France were at war, and within a short while of leaving port Howard's vessel was captured by an enemy privateer. It was then that the future prison reformer had his own first experience of captivity.

For a week he was kept in a hold, and had to sleep on a handful of damp straw. On reaching France he was thrown into various prisons, and saw for himself the appalling conditions in which the inmates had to exist. He learnt that things were no better back home in England, and on his eventual release he decided to try to end the horrors of prison life.

He began his career as the Champion of the Prisoners, as he was called, by becoming the High Sheriff of Bedfordshire. He attended all the court trials held in the county, and later visited the prisoners as they languished in Bedford gaol.

To his horror he discovered that the gaolers received no fixed wages, but lived mainly by taking money from the prisoners under their charge. This fact sent him on a tour of every prison in England, when he persistently attacked the buildings themselves, the system, and the abuses that were practised.

NO HOPE

In some gaols he came across boys of thirteen and fourteen who were confined in dark and airless cells. Many of the men he spoke to had been imprisoned because of debts they could now never hope to pay. And unless the inmates could afford to buy straw from the warders, they had to lie on the bare, stone ground.

Once he had gathered his evidence, Howard's next task was to see that prison reforms were put through Parliament. He was responsible for several humanitarian Bills being passed, including the Bill of 1774 which ensured that every prison cell was scraped and whitewashed at least once a year, and that the prisoners were properly clothed, fed, and medically cared for.

His activities, however, did not end in England. He went on a three year tour of Europe, visiting gaols in all the major countries. Altogether he travelled more than 13,000 miles, during which time he helped to quell a prison mutiny in which two warders were attacked and killed by a mob of two hundred inmates.

After further adventures aboard a plague-stricken Italian ship—when he loaded and fired the only gun to repel an attack by Barbary pirates—Howard went to Russia. There, in 1790, the great philanthropist died while tending to a young woman who was suffering from an infectious fever.

PRISONERS

In Europe, Howard helped to quell a prison mutiny in which two warders were killed by a mob of two hundred inmates.

HERO of the TOURNY

IN the forenoon of 26th October, 1901, the French barque *Tourny* left the Spanish port of Valencia, bound for Marseilles.

As it happened, before the barque had left the Bay of Valencia, she ran into heavy weather. The ballast in her hold shifted until she was lying on her beam-ends.

There was nothing the crew could do to right her and, after a terrifying night, an emergency meeting was held.

From the way she was listing, it seemed that the *Tourny* would sink at any moment. No one could offer any solution, so the captain reluctantly gave the order to abandon ship.

Fortunately a French steamer, the *Italie*, sighted the barque shortly after 10 a.m., and a few minutes later the captain of the *Tourny*, his wife and the vessel's crew of fourteen prepared to transfer to the rescue vessel.

One man, however, doggedly refused to leave his post.

He was Alexis Denis, an ordinary able seaman who faced the captain and said boldly, "I think you are wrong. The ship won't sink. If you won't stay with her, then I will. And I'll bring her safely to port".

By now the captain was convinced that Denis was crazy. He shook the seaman firmly by the hand, believing that he would never see him alive again. The crew then put off in a longboat, and their last sight of Denis was of a "madman" holding on to the bulwarks of the listing ship while his dog raised its head and howled.

A short while later, the *Italie* steamed away, and as the *Tourny* became a smudge on the horizon,

observers on the steamer thought they saw her fill and sink. That, at least, was the news given by the captain to a Marseilles newspaper.

Denis was presumed to be dead, and the paper reported how he gallantly went down with his ship.

Denis, was however, far from being dead. He was very much alive, and the new "master" of the *Tourny*.

For the following two or three days, he waited patiently for the bad weather to subside.

Denis and his dog drifted helplessly south, past the island of Ibiza, in the direction of the North African coast. There was little that he could do to steer the ship, and he was completely at the mercy of the wind and the tides.

A week later, as the *Tourny* neared the coast of Morocco, she at last encountered some good weather. The sea was calm and the wind slight when she was sighted by an English steamer, the *Syrian Prince*, bound from London to Malta.

At first the steamer's look-out thought that the barque had been completely abandoned: but then, to his amazement, he spotted the lone figure of Denis.

The captain of the *Syrian Prince*,

Captain Turner, immediately sent a jolly-boat over to the French vessel. One of the steamer's passengers, a man called Wakeman Long, offered his services as interpreter. A short time later he scrambled aboard the barque, was astounded by what he saw and wrote afterwards:

"Her decks had been completely swept clean of every movable article ... the lower topsails were unfurled, but they had been blown to ribbons and drooped in fantastic drapery around her yards."

Long received the greatest shock of all when he saw the weary Denis standing with his faithful dog.

The seaman wore a small black cap and had a muffler around his neck. "The look in his eyes," said Wakeman, "was that of a man who had not slept for nights."

Denis's opening words confirmed this impression.

"I am not well, monsieur," he told Long. "I am very tired."

The Frenchman was then given a sleeping draught, and a tow-rope was secured between the two vessels. The sea was now becoming rough again, and after ten minutes the line broke and the ships were driven apart.

This happened three times in all,

and Captain Turner was on the point of giving up the rescue operations.

It was then that Denis woke up and produced a six inch grass rope which he said would not break. By this time, passing a tow-line had become a very dangerous operation, and on several occasions the masts of the two ships almost touched. In the emergencies, every member of the *Syrian Prince's* crew, and most of her passengers as well, fell to and gave a hand.

They struggled for hours to make the rope fast. At one moment, disaster almost struck.

"I saw the barque's jib-boom just over our heads," said Wakeman Long. "She lifted on a sea, just cleared our funnel, then fouled our main rigging, bringing down the truck, lightning-conductor and back-stay and removing our flag-pole, while the barque's flying jib-booms snapped like matches."

In spite of this, Denis still refused to leave his ship; even when the stronger tow-line was finally secured, he stayed at the *Tourny's* helm and steered the crippled barque towards Algiers.

The following day, the two vessels sailed slowly into the Algiers harbour, with Denis looking "like a wet rag".

"On nearing port," Mr. Long recalled, "he made a last desperate effort to recover himself, and with a feeble hand hoisted the French Tricolor at the peak."

An hour later, the lone hero of the *Tourny* was taken safely ashore, still accompanied by his dog. Before long, the whole of France acclaimed the man who refused to admit defeat.

Paul Rainer

The Destiny of Dandy George

Two English kings were spellbound by the charm of George Villiers. As a result Villiers, "an amazingly ignorant man," was allowed to plan wars, make speeches, and insult foreign Royalty just as he pleased.

WHEN George Villiers was twenty-two years old he had an income of £1 a week. Two years later it was £300 a week. In those two years Villiers, "an amazingly ignorant man" as he has been called, yet a dashingly handsome fellow, had become a darling of fortune, the favourite of King James I.

In those two years, too, the great essayist Francis Bacon wrote to Villiers: "You are a new risen star: let not your own negligence make you fall like a meteor."

Sound advice—and if only Villiers could have profited from it! Instead, supported by no other ability than his good looks, he embarked upon a career of adventurous roguery that won him wealth for which he was poorly suited, notoriety which made him hated—and a terrible death at the point of a dagger.

Villiers was born at Brooksby in Leicestershire in 1592, and was a man of extraordinary charm; tall, slender, with a pointed beard, flashing eyes and a wonderful complexion. Courageous, gallant and just, he was nonetheless completely unskilled in politics— a fact which did not deter him from laying down the law in conversation that sparkled with wit and character.

Clearly, if such a man were ever to meet up with one of the vain, self-confident Stuart Kings of England he was destined for fame. Villiers, to his own everlasting good fortune, met King James I at Newmarket in August, 1614. Quickly, and inevitably, he impressed the King and was appointed a royal cupbearer.

Lord High Admiral

BY the following May, Villiers had been promoted, knighted and had won the King's firm affection. And a year later he was an earl, worth £15,000 a year. Indeed, it became evident to James's astonished courtiers that no gift was thought to be too big by James to give this genial, radiant young man: Lord High Admiral, Chief Justice of the Parks and Forests south of the Trent, Master of the King's Bench office and Constable of Windsor Castle were among the titles

quickly heaped on Dandy George.

In a spirit of true roguery, Villiers used his influence with the King to interfere with the law, advance his own family, demote people who opposed him, flirt outrageously with the noblewomen at the court—and even boasted that he was "Parliament-proof!"

Then came a strange adventure that shows off his character. The King was anxious to get his son, the future Charles I, married to Maria of Spain. Accordingly, Villiers and Prince Charles set out as ordinary travellers for Spain: an opportune moment for Villiers to leave England, for he was already becoming most unpopular.

In Spain the Prince and his companion were received with the highest honours, so high indeed that King James wrote to them: "The newes of youre gloriouse reception thaire makes me afrayed that ye both miskenne your old dade hereafter"—the last part of which suggests that the Prince might forget his old Dad!

But like so many things in which Villiers had a hand, the mission was unsuccessful, partly because both sides disagreed about the marriage terms and partly because the two envoys did not make themselves liked. The Spanish certainly did not like Villiers's familiarity with his Prince—they did not understand a nobleman who talked to his prince in shirtsleeves.

Villiers, having failed, became bitterly opposed to the Spanish marriage. King James did not seem to mind a

The naval officer's dagger clattered to the floor—and George Villiers fell to the ground.

bit, however, for when Villiers returned to England the King embraced him—and then made him the Duke of Buckingham.

The new Duke's new plans were for a war with Spain and for Prince Charles now to marry a French princess. The marriage came after James died, on March 27, 1625, leaving his successor still under the domination of the favourite.

In 1626, when his friend Charles went to his Coronation, Villiers offered the King his arm as a support while he mounted the steps of the dais. Charles was heard to say: "I have more need to help you than you have to help me." That same year the Commons demanded Villiers's dismissal, whereupon the King promptly dissolved Parliament.

Conspicuous Gallantry

AS English relations with France worsened it was decided to send an expedition in support of the French Huguenots (Protestants). In June, 1627, Villiers sailed with six or seven thousand men and landed on the Isle de Rhé. After a vain attempt to storm a French fortress he was forced to retreat. He behaved with conspicuous gallantry, and was the last person to leave the beach, but this little trip cost nearly

four thousand casualties, and when in November he landed in England he was well on the way to becoming the most hated man in the kingdom.

It seemed that Villiers would never learn, for he went right on planning attacks on the French. On August 17, 1628, he posted off to Portsmouth to hurry forward his next expedition. The outcry against him in that city was considerable: in the streets an attack was made on him by drunken or mutinous sailors. His doctor, whose name was Lamb, fell a victim to the mob instead, and after that sinister rhymes pursued Dandy George:

"Let Charles and George do what they can
The Duke shall die like Dr. Lamb."

The events of the next few days were dramatic indeed. On August 22 Villiers was unwell and the King came to see him at the house in which he was lodging at Portsmouth. When the King left Villiers embraced him in a most unusual manner. Early the next morning, just as he had finished breakfast, he came out into the hall of his house. A fanatical naval officer, named Felton, was waiting there, and instantly struck him in the heart with a dagger.

George Villiers, Duke of Buckingham, staggered and fell down, saying, "God's wounds, the villain hath killed me!" A few minutes later he was dead.

NARVIK

ATLANTIC OCEAN

NORWAY

SWEDEN

FINLAND

GULF OF BOTHNIA

OSLO

STOCKHOLM

BALTIC SEA

U.S.S.R.

VICTORY AT A PRICE

In the Second World War Narvik was a German victory—but a costly one. In it the small German navy suffered crippling losses from which it never effectively recovered

Guns blazed as for the third time English destroyers attacked German ships in Narvik harbour.

ON the bridge of his destroyer, Rear Admiral Bonte stared anxiously at the hard outline of the Norwegian mountains thrown into dark relief by the rising sun. It was very cold, and the Admiral blew into his cupped hands to warm them.

Behind his flagship, the nine other destroyers of the German flotilla took up their battle stations. In front, under the angry brow of the mountains, lay the port of Narvik, a vital link in the rail and sea communications of Scandinavia.

This spring of 1940 was no time to stand and applaud nature's wonderful reawakening. For at this moment German troops were marching into Denmark and German warships were steaming around the Norwegian coast, ready to support landings. To Rear Admiral Bonte, with his ten destroyers, Hitler had given the task of capturing Narvik.

"I rely on you, Admiral," he had said, "to secure us a base on Norwegian territory. It is an important task, but I do not think that you will fail me."

The Admiral had saluted and hurried off. Now, standing on the bridge of his flagship, he asked himself the all-important question: "Will the Norwegians resist?"

A shot rang out, and a shell whistled overhead and exploded harmlessly in the water between the ships. From the Norwegian warship Eidsvold, which was just visible in the dawn, came the flash of an Aldis lamp.

"They are asking us to identify ourselves, Admiral," said the destroyer's captain. Bonte thought carefully for a few moments, then sent an officer in a motor-boat to ask for surrender.

"We will not surrender. We will resist," said the Norwegian commander stiffly, when the young officer repeated Bonte's demand. The German saluted and hurried back to his launch, from which he signalled to the flotilla.

BRITISH ATTACKS

BONTE acted swiftly. As soon as the launch was out of the line of fire, his destroyer swung round and fired torpedoes that blew the Eidsvold to pieces. At that instant, another Norwegian warship, the Norge, opened fire, but she too, was sunk.

Quickly, the German flotilla edged to the quayside, and disembarked two battalions of troops with their equipment. They were commanded by an officer who was later to play an important part in the battle, Brigadier-General Dietld.

By eight o'clock, the port was in German hands. Well pleased with his success, Admiral Bonte radioed his report and went below to breakfast.

But if he thought that the fighting was over, the Admiral was mistaken.

The following day five British destroyers steamed into the port and in a savage battle sank two of the five German warships that were in the harbour at that time. One of the German seamen who died in that brief action was Rear Admiral Bonte. He had had only twenty-four hours in which to enjoy his victory.

The British force turned away, their work done. The heaving waters of the port were covered with floating debris, clothing, cordage, all the flotsam of battle. But even as they drew out of the port, the destroyers were attacked by the five other German warships that had been attracted from their anchorages by the gunfire. The two sides were equally matched, but the German destroyers had heavier guns, and one British destroyer was sunk, another ran aground and a third was damaged.

Even this second pitched battle did not discourage the Royal Navy.

With characteristic determination they managed to sink a large German cargo ship as she put about to enter harbour.

"Retire to the west," ordered Sub-Lieutenant Davies, as he read that signal from the leading destroyer. "In other words, don't press your luck too far!"

His companions grinned.

Another day passed. Then another. Once more the Germans settled down and relaxed their watch. Surely even the stubborn British could recognize defeat? But again they were wrong, for on April 13, the battleship Warspite sailed in from the west, leading a new destroyer flotilla. Their commander, Vice-Admiral Whitworth, pressed home the attack. There was a brief, angry exchange of fire, and then the flotilla withdrew from the battle leaving the German destroyers lying at the bottom of the harbour.

Inland, Brigadier-General Dietld received the news with dismay. This short, vigorous man now realized that he would have to maintain his two battalions of troops in the hostile mountains until help could be sent. Yet, with the cunning that later earned him the nickname of the "Arctic Fox," he managed to keep in the fight.

END OF THE BATTLE

FOR the next few weeks he had much to worry about. On May 28, 25,000 Allied troops drove him from the country around Narvik and farther up into the mountains, towards the Swedish frontier. The situation became desperate. Elsewhere in Norway, German forces were suffering great losses: the brand-new 10,000 ton cruiser Blücher was sunk off the coast with all hands, and a pocket-battleship was badly damaged.

In those weeks of fighting, when the Royal Navy fought desperately to hold the enemy, the Germans lost ten destroyers and three cruisers at Narvik. Two great battle-cruisers were so badly hit that it was many months before they were fit to go to sea again.

But time was against the Allies. On May 10, Hitler flung his armies against France, Belgium and the Low Countries, and the 25,000 men who had driven Brigadier-General Dietld inland were withdrawn. Narvik was abandoned.

The "Arctic Fox" stole out of hiding and recaptured the port. Later, he was promoted and awarded the Knight's Cross of the Iron Cross, a high military decoration.

"Dietld," Hitler proclaimed, "is the Victor of Narvik!"

Narvik certainly was a German victory, but it was a costly one. Many warships had been lost in the battle, and the small German Navy could not stand such losses. In the Battle of Narvik, it suffered a blow from which it never recovered.

SIGNS & SYMBOLS
—a History of Handwriting

WHENEVER we write something down, we use either letters of the alphabet or numbers. Occasionally, we use other symbols. For instance, one of the most common symbols which is neither alphabetical nor numerical is the sign &.

Everyone knows what it means: it represents "and," and is usually seen in the names of firms, such as Jones & Co., Smith & Brown, and so on. This sign has a special name; it is called "ampersand." Ampersand is used by all people who employ the Latin alphabet, and no matter what language is involved, the sign always means the same: "and."

Really, it is made up of two letters; a cursive capital letter E and a small T. This can be seen more easily in an italic &. The E and T make up the word "et" which is Latin for "and," and you can realise this better when ampersand is used in an alternative abbreviation for "etcetra," which can be shown as "Etc," or "&c."

The name "ampersand" itself comes from a corruption of the actual description of the symbol, which is "and *per se* and." *Per se* is a Latin phrase meaning "by itself," so the description really means "and by itself and."

Another well-known symbol is our familiar £ sign, which is really a script capital L with two strokes across it. The L is short for *Libra*, the Latin word for pound. The abbreviation "lb" is also derived from "libra," or "librae" in the plural (which shows why it is wrong to write "lbs" for the plural).

Two other familiar symbols which come from the Latin are the question mark (?) and the exclamation mark (!). The question mark is taken from the abbreviation of the Latin word *Quaestio*, question, written as a letter Q above a small $_0$, thus: Q which gradually became converted into our symbol. The exclamation mark is formed from a Latin word **Io** signifying "hurrah." This too was written with one letter above the other, thus: I and, after a while, the I

became converted into a straight line, and the $_0$ into a dot.

The Americans have a sign comparable with our £ for denoting dollars, the well-known $. No one really knows for sure how this sign came into existence, but the most acceptable theory is that it is adapted from the symbol which appeared on the old Spanish dollars, which was a figure "8." These coins were known as "pieces of eight" and, when writing accounts, a large figure "8" was used, with either one or two cancelling lines drawn through the figure. The symbol was gradually used for American dollars, and became transformed into the present one.

In Britain, we often express prices in shillings and pence by drawing a slanting line between them, such as 5/6d. for five shillings and sixpence. Few people ever consider why we do this. In fact, it dates back to the time when people used the long, flowing S, thus: $\int$. Five shillings and sixpence would then be written 5 $\int$ 6d.; and the flowing "S" gradually became a line.

By far the most common symbols are the letters of the alphabet and the figures one to nine and zero. The figures are usually called Arabic numerals, to distinguish them from the older Roman numerals. Nowadays, Roman numerals are only used for such things as clock faces and inscriptions carved into buildings. But in Roman times they were the only way of writing down numbers.

Now we come to those familiar symbols which make up our ABC. So far as we know, the alphabet used in Europe and the Mediterranean countries (and now spread all over the world) started with the Ancient Egyptians, who used a picture writing system.

The ancient Egyptians used their picture symbols in two ways: the first was quite simple. They simply used a picture of the object to denote that object. For instance, in English, we might use a picture of a bee to denote a bee, without any words at all, and we would be quite easily understood.

The second way was to use the symbols phonetically; that is, by making use of the *sound* of the word, quite independently of what it means. Suppose we take our bee symbol again, in English we might use it together with a picture, say, of a leaf:

Drawn together on a page, we would look at the two symbols, and pronouncing their names one after the other, we would get "bee-leaf." Thus, we would have expressed the word "belief" phonetically by using picture symbols.

The ancient Egyptians did exactly this, except, of course, that they used the ancient Egyptian language instead of English. To take an example again, the word in ancient Egyptian for "water" was *mu*. Now the symbol means "owl," which is what it looks like. The symbol means "rope." But the "owl" symbol also represents the sound M, and the "rope" symbol also represents the sound U. Put together, thus: the two symbols represent MU, meaning water.

In this way, an alphabet was formed, but it is obvious that to use such complicated symbols, such as having to draw a little picture of an owl every time one wanted the letter M, would make life very difficult for scribes and letter-writers. So a second, more simple, alphabet was designed. The picture symbols are called the "hieroglyphic" alphabet, and the simpler signs are known as the "hieratic" alphabet. The picture symbols were normally used for inscriptions on buildings and tombs, and for the

more important documents, while the simpler form was in everyday use.

For some people, even the simplified "hieratic" alphabet was too complicated, and among these people were the Phoenicians. They were a nation which lived in what is now Syria and the Lebanon. The Phoenicians were great traders, and they sailed their ships all over the Mediterranean and even beyond, reaching the shores of Britain. It was of enormous importance that they should be able to write things down, and so they adapted the hieratic Egyptian alphabet to their own use, but made it more simple and angular.

As the Phoenicians travelled, they took their methods of writing with them, and so the idea of an alphabet was spread. The Romans and Greeks adopted versions of the Phoenician letters, while the Hebrews invented rather more square characters, which made them look less like the Phoenician originals.

As time went on the shapes of

Left:
A diagram showing the development of numerals.
Below:
A table of letters from various languages.

A diagram showing the development of numerals.

	Fourteenth Century European	Twelfth Century European	Twelfth Century Arabic	Tenth Century Indian	Modern Arabic	Modern European
1						1
2						2
3						3
4						4
5						5
6						6
7						7
8						8
9						9
0						0

A table of letters from various languages.

English	Russian		Greek		Hebrew
A	А	а	А	α	א
B	Б	б	В	β	ב
V	В	в			ו
G	Г	г	Γ	γ	ג
D	Д	д	Δ	δ	ד
E	Е	е	Е	ε	
Zh	Ж	ж			
E			Н	η	
I	И	и			
I	Й	й	Ι	ι	י
H					ה
K	К	к	К	κ	ק כ
L	Л	л	Λ	λ	ל
M	М	м	М	μ	מ
N	Н	н	Ν	ν	נ
O	О	о	О	ο	
P	П	п	Π	π	פ
R	Р	р	Ρ	ρ	ר
S	С	с	Σ	σ or ς	ס שׂ
T	Т	т	Т	τ	ט
U	У	у	Υ	υ	
F,Ph	Ф	ф	Φ	φ	
Kh,Ch	Х	х	Х	χ	ח
Ts	Ц	ц			צ
Ch	Ч	ч			
Sh	Ш	ш			שׁ
Shch	Щ	щ			
	Ъ	ъ			
X			Ξ	ξ	
Y	Ы	ы			
	Ь	ь			ע
E	Э	э			
Yu	Ю	ю			
Ya	Я	я			
Z	З	з	Ζ	ζ	ז
Th			Θ	θ	ת
Ps			Ψ	ψ	
O			ω or Ω	ω	

19

the letters changed, until they reached their present forms. The alphabet we use today is called the Latin alphabet to distinguish it from other alphabets, such as the Greek, Hebrew and Russian.

There are usually 26 letters, although the Romans had only 22 of them, for they had no G, J, U or W. All these have been invented since. In fact, the English alphabet used to have three more characters which are no longer used. One was called *thorn*, and looked like this: Þ. It represented the sound TH in the words "the" and "that." Another symbol was called *wen*, sounding like "w," which was written: Þ. Later still, the letter D was adapted for the sound TH in words like *thing* and *thick*. This symbol was written as a capital: Ð or ð as a lower case letter. After a while, these letters were dropped, except in words like *the* and *that*, which were still written Þe and Þat

Then people got these symbols confused with Y, and so wrote *Ye* and *Yat* for *the* and *that*. We can still see examples of this confusion in signs on country shops, when they bear inscriptions such as "Ye Olde Tea Shoppe." Of course, the word *the* was never spelled *ye*. It is simply a case of mixing up the old letters.

Both ð and Þ are still used today in the Icelandic language, which otherwise uses standard Latin characters.

The Latin alphabet has three close relatives: the Greek, Russian and Hebrew. In Britain we seldom come across Russian letters, except on stamps from the USSR and other Slav countries, but Greek and Hebrew are less uncommon.

Jewish shops, especially butchers, often have the sign כשר displayed, which means "kosher." This tells customers that the food sold has been prepared according to strict Jewish law. The Hebrew alphabet is read from right to left. The ancient Egyptians, and several other nations, used to write both from left to right, and from right to left, so the Hebrew alphabet is not so unusual as it seems.

In the sign כשר , reading from right to left, the letters are: כ , *kaph*, sounded as K, שׁ *shin*, sounded as SH, and ר *resh*, sounded as R. Vowel sounds are indicated by dots inside the letters, or else beside them.

Another alphabet sometimes met with in Britain is the Greek, which again can be seen on shop windows belonging to Greek and Cypriot immigrants. Also, there are Greek language newspapers and books published in Britain which use this alphabet. The modern Greek alphabet is practically the same as

ABCDEFGHILMNOPRSTU
abcdefghilmnopꞃꞃtu

An example of the Irish Alphabet.

ABCDEFGHJKLMNOP
QRSTUVWXYZ
abcdefghijklmnopqrsſtuvwxyz

An example of the German Black Letter.

The illustration above is of Jean Mielot, a French calligraphist of the fifteenth century.

that used by the ancient Greeks. Like the Jews, the Greeks gave names to their letters, and in fact the word "alphabet" is named from the first two letters of the Greek alphabet, which are: A *alpha*, B *beta*.

If you ever visit London Airport you will probably see a Russian airliner, which will have inscriptions on its side in Russian characters. Looking at such inscriptions, the average Englishman would probably dismiss them as quite incomprehensible. At first glance it almost looks as if someone was having a little joke, by taking the ordinary Latin characters, turning some back to front and adding others which are complete inventions.

Yet a closer look at the Russian alphabet will show that there are some letters which look and are pronounced just like our own.

A, E, K, M, O, T are all pronounced in roughly the same way in both alphabets. But there are some others, which although they *look* like our letters, are really quite different.

B in Russian is in fact a V, H is really an N, P is sounded like R, C is the Russian way of writing S, while X is a sound like "ch" in Scottish *loch*.

The sign И, like a backwards N,

"அம்மா! நான் கிட்டப்போனால் வெள்ளை மாடு முட்ட வருகிறது. கருப்பன் போனால் ஒன்றும் செய்யாமல் சும்மா நிற்கிறதே. இது ஏன் இப்படி ?"

Tamil India

От заник-слънце озарени,
Алеят морски ширини;
В игра стихийна уморени,
Почиват яростни вълни...

Bulgarian

Μῆνιν ἄειδε, θεά, Πηληιάδεω Ἀχιλῆος
οὐλομένην, ἣ μυρί' Ἀχαιοῖς ἄλγε' ἔθηκεν.

Greek

Sgiać ċoranca na fípinne calma
an ɼiormapnać bɼéiʒe
Buan-ɼcoc na Ceannaioċta

Irish

ⰔⰑⰂⰡⰕⰟ ⰐⰑⰂⰟⰋ Ⰳ ⰏⰑⰓⰡⰐⰋⰆⰠ
ⰒⰀⰓⰠ ⰒⰑ ⰓⰑⰏⰠ ⰞⰀⰓⰟⰎⰟⰀⰆⰞ

Glagolti (old Slav.)

একদিন অপেক্ষাকৃত অল্পবয়সে যখন আমার শক্তি ছিল তখন কখনো কখনো ইংরেজি সাহিত্য মুখে মুখে বাংলা করে শুনিয়েছি আমার শ্রোতারা ইংরেজি জানতেন সবাই।

Bengali

ארשם על נחתחור כנזסרם וכנותה וכעת
פֿכוסרי שמה ורשבר עלים

Hebrew

أَلْحَمْدُ لِلَّهِ رَبِّ ٱلْعَالَمِينَ ۞ ٱلرَّحْمَٰنِ ٱلرَّحِيمِ ۞ مَالِكِ يَوْمِ ٱلدِّينِ ۞ إِيَّاكَ نَعْبُدُ وَإِيَّاكَ نَسْتَعِينُ ۞ ٱهْدِنَا ٱلصِّرَاطَ ٱلْمُسْتَقِيمَ ۞ صِرَاطَ ٱلَّذِينَ أَنْعَمْتَ عَلَيْهِمْ غَيْرِ

Arabic

იყო ერთი ჭურჭია მჴეჭვალი. რაც უნდა გამოეჭრა. უთუღათა ნაჭერი უნდა მოეჭარა. ერთი ლამის ნახა სიზმარი, ვითომცა ჰირმი ამოსლოდა დიდი

Georgian (U.S.S.R.)

ℒ glücklich, wer noch hoffen kann,
Aus diesem Meer des Irrthums aufzutauchen!
Was man nicht weiß, das eben brauchte man,

German Black Letter

સારો પવન જોઈને પછી અમે બોર્નિઓ નેટમાંથી નીકળ્યા. પણ આગળ ચાલતાં તોફાન લાગવા માંડ્યું. કેટલાક દહાડા લગી તો અમારું વહાણ ઊંછળતુંજ રહ્યું. છેવટે પવન નરમ પડ્યો અને

Gujerathie (India)

is really I, while an R back to front is sounded YA!

The Cyrillic alphabet, which is the proper name for the Russian alphabet, is in general use throughout the U.S.S.R., although there are some other alphabets in use there. It is also used, with slight variations, in Bulgaria and Jugoslavia.

The Serbs and Croats of Jugoslavia speak a language which is practically identical, but because of an accident of history, the Serbs write their language, referred to as Serbian, in Cyrillic characters, while the Croats write their version, or Croatian, in Latin letters. This explains why signs in Jugoslavia, as well as their money and stamps, are designed with inscriptions in both alphabets.

The present Russian alphabet has 32 letters. At one time there were even more, but after the Russian Revolution in 1917 the alphabet was revised to its present form.

The name Cyrillic is given to this alphabet because it is reputed to have been invented by a monk named St. Cyril (also known as Constantine the Philosopher). St. Cyril worked on an alphabet about the year 855 A.D., and produced a total of 38 letters, based largely on the Greek characters, but with additions for sounds which were peculiar to the Slav tongues.

During the Middle Ages, a form of the Latin letters developed, known as the "Black Letter," or more popularly as it is known today, "Old English." This died out in England in the fifteenth century, to be replaced by the Roman and Italic styles which had spread from Italy. This style has remained almost unchanged until the present day.

In Germany, however, the Black Letter style continued, and it was only in recent years that it was generally replaced by the more common and readable Roman style.

Even before the Latin alphabet reached the shores of Britain, the Saxons were already using an alphabet of their own. This was called Runic, which had been in use for centuries. Old inscriptions can still be found in England, Ireland, Germany and Scandinavia. Each of these alphabets varied slightly, but there were 16 basic letters. Seven more were added later.

Even earlier, and peculiar to the British Isles only, was an alphabet called Ogam, which appeared as notches carved on the corners of wooden staves or on square-shaped stones.

Today, there are upwards of fifty alphabets in daily use all over the world. These do not include such inventions as shorthand, nor can the Chinese system be called an alphabet, since Chinese is pure picture-writing, each character representing a word. Although somewhat cumbersome, it has its advantages, since there are many languages spoken in China, most of which are quite different from each other.

The Japanese have adapted the Chinese characters to their own use, and have formed a *syllabary*. In this method, each character is used to represent a *syllable*.

21

"Having taken in water, we sailed forward five days near the land until we came to a large bay which our interpreters informed us was called the Western Horn.   In this was a large island, and in the island a salt-water lake, and in this another island, where, when we had landed, we could discover nothing in the daytime except trees ; but at night we saw many fires burning, and heard the sound of pipes, cymbals, drums and confused shouts.   We were afraid, and our diviners ordered us to abandon the island . . ."

The

Great Voyage

THE words on the facing page are taken from the oldest account in the world of a voyage of discovery. They are attributed to a Phoenician named Hanno who led a fleet of ships out of the Mediterranean and down the west coast of Africa in about 500 B.C.

On his return to Carthage, the Phoenician capital on the coast of North Africa, Hanno had a tablet, inscribed with an account of his travels in the Phoenician language, hung in the Temple of Melkarth. The Carthaginians' great enemies, the Romans, later destroyed Carthage, but luckily the account had been translated into Greek before this happened.

According to the tablet, Hanno was ordered to take a fleet of 60 ships out past the Pillars of Hercules (as the Straits of Gibraltar was called in those days), and establish colonies on the Atlantic coast of Africa. Each ship could carry five hundred people and was powered by fifty huge oars.

It is believed that Hanno's description of his route was left deliberately vague to avoid giving anything away to Carthage's trading rivals; but in modern times his route has been plotted from his description of the coastline. It is thought that he covered 3,000 miles from the Straits of Gibraltar and reached the area which was later called Sierra Leone.

According to the account on the tablet, Hanno established his first colony two days after he had passed through the Pillars of Hercules. This he named Thymiaterium. Altogether he established six colonies.

At length the fleet came within sight of the Anti-Atlas Mountains. Anchoring at a river mouth, the voyagers made friends with the local natives. Some agreed to sail with Hanno as interpreters.

The voyage continued, the huge galleys coasting past mountains and desert land until they reached a deep and wide river which Hanno notes was filled with "crocodiles and water-horses". It is believed that at this point he was off the coast of Senegal.

After this, the fleet, its great oars dipping into the Atlantic swell while the slaves sweated on the rowers' benches, reached the archipelago between the River Jeba and Cape Verga and anchored off an island which was probably situated in the estuary of the River Pongo. It was here that the Carthaginians heard the pipes, drums and cymbals at night and, being afraid of the unknown, sailed hastily on.

Unfortunately Hanno's account of the voyage is very brief. It tells us nothing of Hanno himself (we know he must have been a very daring commander to take his ships so far into the unknown), and much fascinating adventure is dismissed in a few words. For example, after leaving the island, the account continues:

" . . . Passing on for four days, we discovered at night a country full of fire. In the middle was a lofty fire, larger than the rest, which seemed to touch the stars. When day came, it turned out to be a large hill called the Chariot of the Gods."

Fire Danger

It is obvious from this description that the fleet passed a volcano in eruption. One can imagine how anxious the sailors were to get past, with the ash raining down on the galleys and perhaps even causing their sails to smoulder. At the same time they had to hug the coast for the fear that, if they passed out of sight of land, they would be swept over the edge of the world.

The "country full of fire" has been identified as Sierra Leone, and the wider fire was no doubt a grass and bush fire.

By this time the fleet must have been running short of supplies, especially as the mangrove swamps which fringed the coast made landing from the big galleys almost impossible. But Hanno still continued, and records that, on the third day after seeing the Chariot of the Gods:

" . . . we arrived at a bay called the Southern Horn, at the bottom of which lay an island like the former, having a lake and in this lake another island, full of savage people, the greater part of whom were women whose bodies were hairy and whom our interpreters called 'Gorillae'. Though we pursued the men, we could not seize any of them, but all fled from us, swinging themselves down from above and defending themselves with stones."

Modern experts think that these strange, hairy "people" were not gorillas but baboons. Baboons still live in that region, and will defend themselves by stone-throwing. The females also carry their babies on the backs rather like native women, which may have caused Hanno to note that many were "women."

Hanno's men finally managed to catch three of these females, but they fought and scratched and bit so much that the sailors killed them. They then skinned them and took the hairy hides back to Carthage, to the wonder of the people there.

At this point Hanno's account ends, with the brief explanation that because of shortage of supplies he was forced to turn his ships round and head back to Carthage.

Two thousand years were to pass before any other explorer followed Hanno's route, when the Portuguese began to edge along the mysterious coastline of West Africa. Even then it took them a century to sail as far as the point reached by the daring Carthaginian.

The History Picture Quiz

Now that you have read the History section, see if you can answer the simple picture-quiz.
The answers are at the back of the book.

1 Where and in what year were the first of the modern Olympic games held?

2 Who almost won the 1908 marathon at the Olympiad in London and why was he disqualified?

3 Who was the negro Olympic sprinter who angered Hitler, and how many gold medals did he win?

4 This ship never sailed. What kind of ship is it, what is its name and what happened when it was launched?

5 Who sailed to Greenland in the ship, *Fram*?

6 When a famous Norwegian explorer was icebound in Godthaab, he lived a hard life. But with whom?

7 What was the real name of this man known to all as Dandy George?

8 British ships are attacking German vessels during the Second World War. In which harbour did this take place?

9 Where would you have to go to find people who wrote characters like this?

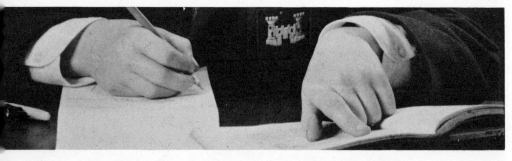

10 What is the correct name for the Russian alphabet? How many letters does it contain?

11 What does the Hebrew sign outside Jewish butchers' shops mean? Is it read from left to right?

12 What are the name and meaning of this?

THE AGE OF FLIGHT

FOR many thousands of years, the power to fly was considered a magical and mysterious gift. Then, men who could think and build, and some who had great daring, began to try to copy the birds. But not until 1783, and the Montgolfier hot air balloon, could men stay in the air; but they could not control their direction. Next came airships, which were also at the mercy of the wind and weather.

As the science of aeronautics progressed, gliding flights were made, but not until the 1890s did the first glimmerings of *control* appear. At the same time, in many countries, came powered "hops" by machines with a variety of engines.

Then came the magic day of December 17th, 1903. On the windswept Kill Devil Hills near Kitty-Hawk, North Carolina, U.S.A., the brothers Orville and Wilbur Wright made the first controlled, sustained, powered flight, after years of experiment and glider flying. The best of these first flights covered 852 feet and lasted 59 seconds.

The first *practical* flying machine was their Wright Flyer III (1) of 1905 which could turn and circle under full control and could stay in the air for more than half an hour.

The science of flight rushed on. Another great milestone came with the Handley Page H.P. 39 (2) of 1931. Designed for an air safety competition, it was fitted with "slots" and "flaps" at the front and rear edges of its wings which enabled it to fly near the ground at astonishingly low speeds in complete safety. The much developed modern versions of these devices are fitted to most big jet airliners.

The two other aircraft illustrated are the newest products of aviation science and engineering. The Grumman F-14A (3) two-seat fighter combines 1,500 m.p.h. speeds with a turn-on-a-sixpence agility. Its "swing-wings" fold back to a thin arrowhead (as shown) for 1,500 m.p.h., and swing out straight to allow it to land at less than one-tenth of that speed.

The McDonnell-Douglas DC-10 (4) is one of the new breed of "wide body" airliners. It is capable of carrying from 255-343 passengers in great comfort, cruising at 575 m.p.h. for 3,200 miles.

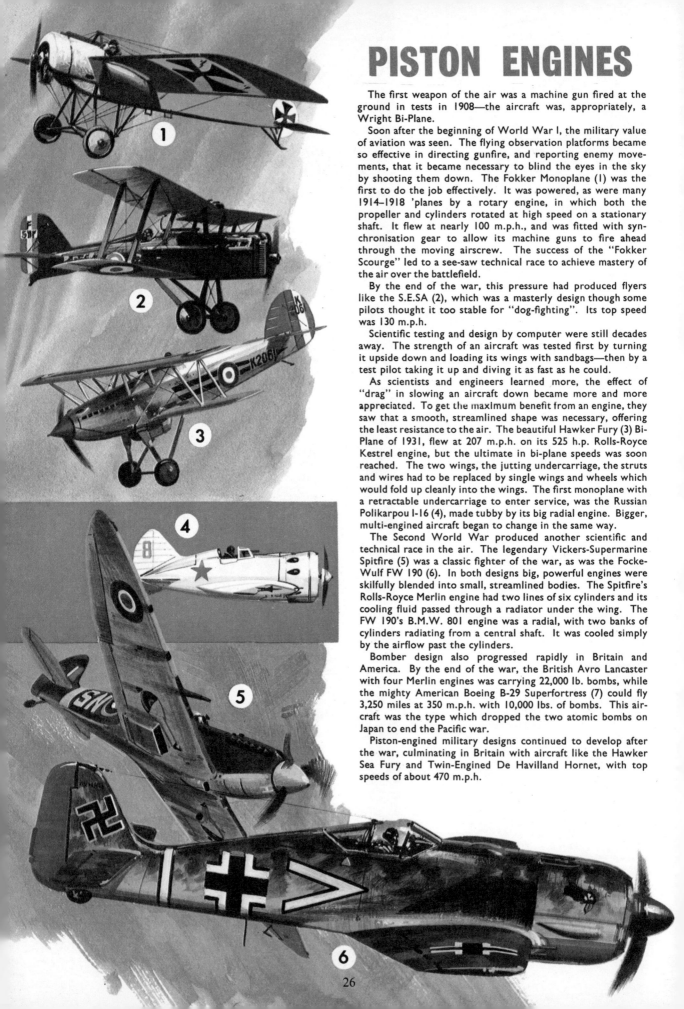

PISTON ENGINES

The first weapon of the air was a machine gun fired at the ground in tests in 1908—the aircraft was, appropriately, a Wright Bi-Plane.

Soon after the beginning of World War I, the military value of aviation was seen. The flying observation platforms became so effective in directing gunfire, and reporting enemy movements, that it became necessary to blind the eyes in the sky by shooting them down. The Fokker Monoplane (1) was the first to do the job effectively. It was powered, as were many 1914–1918 'planes by a rotary engine, in which both the propeller and cylinders rotated at high speed on a stationary shaft. It flew at nearly 100 m.p.h., and was fitted with synchronisation gear to allow its machine guns to fire ahead through the moving airscrew. The success of the "Fokker Scourge" led to a see-saw technical race to achieve mastery of the air over the battlefield.

By the end of the war, this pressure had produced flyers like the S.E.5A (2), which was a masterly design though some pilots thought it too stable for "dog-fighting". Its top speed was 130 m.p.h.

Scientific testing and design by computer were still decades away. The strength of an aircraft was tested first by turning it upside down and loading its wings with sandbags—then by a test pilot taking it up and diving it as fast as he could.

As scientists and engineers learned more, the effect of "drag" in slowing an aircraft down became more and more appreciated. To get the maximum benefit from an engine, they saw that a smooth, streamlined shape was necessary, offering the least resistance to the air. The beautiful Hawker Fury (3) Bi-Plane of 1931, flew at 207 m.p.h. on its 525 h.p. Rolls-Royce Kestrel engine, but the ultimate in bi-plane speeds was soon reached. The two wings, the jutting undercarriage, the struts and wires had to be replaced by single wings and wheels which would fold up cleanly into the wings. The first monoplane with a retractable undercarriage to enter service, was the Russian Polikarpou I-16 (4), made tubby by its big radial engine. Bigger, multi-engined aircraft began to change in the same way.

The Second World War produced another scientific and technical race in the air. The legendary Vickers-Supermarine Spitfire (5) was a classic fighter of the war, as was the Focke-Wulf FW 190 (6). In both designs big, powerful engines were skilfully blended into small, streamlined bodies. The Spitfire's Rolls-Royce Merlin engine had two lines of six cylinders and its cooling fluid passed through a radiator under the wing. The FW 190's B.M.W. 801 engine was a radial, with two banks of cylinders radiating from a central shaft. It was cooled simply by the airflow past the cylinders.

Bomber design also progressed rapidly in Britain and America. By the end of the war, the British Avro Lancaster with four Merlin engines was carrying 22,000 lb. bombs, while the mighty American Boeing B-29 Superfortress (7) could fly 3,250 miles at 350 m.p.h. with 10,000 lbs. of bombs. This aircraft was the type which dropped the two atomic bombs on Japan to end the Pacific war.

Piston-engined military designs continued to develop after the war, culminating in Britain with aircraft like the Hawker Sea Fury and Twin-Engined De Havilland Hornet, with top speeds of about 470 m.p.h.

Civil air transport began in earnest after World War I, with small, brave companies operating converted warplanes like the De Havilland D.H.9 Bi-Plane, which carried two passengers and a pilot adventurously to Paris. In America hardened pioneers like Undbergh carried the mail in the teeth of the blizzard and fog. The designers were already working to take the adventure out of civil flying. The public became gradually more air-minded thanks to the courage of exploits like Alcock and Brown's 1919 Trans-atlantic Flight, Undbergh's 1927 New York-Paris Solo, Kingsford-Smith and Ulm's 1928 Trans-Pacific Flight, Amy Johnson's 1930 England-Australia Solo and the work of men like Sir Alan Cobham, who carved out air routes across the world.

The Handley Page H.P.42 (8) of the early 1930s was safe and steady as a house. In years of flying to the Continent, the type never had an accident. The bi-plane airliner went the way of the military bi-plane and new shapes were in the air.

The world-famous Douglas Commercial (DC) designs tell the story of piston-engined monplane airliners in themselves. In 1933, came the DC-1, a 14-passenger all-metal twin-engine design. 1934 brought the improved DC-2 and in 1935 the world's most famous transport aircraft, the DC-3 (9), known by its RAF name "Dakota". It brought real luxury to the airlines, and in World War II, 10,000 were turned out as cargo and troop-carriers and glider tugs. The DC-3 is still flying all over the world.

In 1946, came the four-engined DC-4, begun as a war transport but switching to the airlines in peacetime. Alongside it, flew the graceful Lockheed Constellation and the portly Boeing Stratocruiser. The "stretching" of the DC-4, produced the 1947 DC-6, and then the beautiful ultimate in piston-engined airliners, the 1953 DC-7C (10). The "family resemblance" between the DC-3 and the DC-7C can be clearly seen, spanning 18 years of technical development.

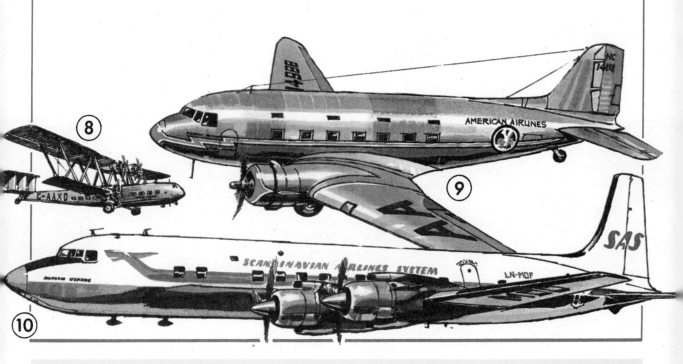

THE THIRTY-YEAR RECORD

On April 26, 1939, Flugkapitan Fritz Wendel, flying a Messerschmitt Bf 109R, set up a piston-engined speed record of 469·2 m.p.h. That record stood unbroken until August 16, 1969, when an American test pilot Darryl Greenamyer, flying his own much-modified Grumman F8F-2 Bearcat (12) beat it by achieving a mean speed of 477·98 m.p.h.

JET PROPULSION

At speeds of 450-500 m.p.h., the propeller begins to lose efficiency. Long before propeller-driven aircraft had reached these speeds, far-sighted men of science and engineering genius were exploring new methods of propulsion in England as early as 1928. A young Royal Air Force officer named Frank Whittle was working on his ideas for a gas turbine engine and, in Germany, Pabst von Ohain was working quite separately in the same direction.

By 1935, Whittle had enough financial backing to begin building and testing his engines—but senior Government scientists could not see any serious future for the invention! In Germany, there was more enthusiasm—by 1939, a jet-powered aircraft existed, the Heinkel He 178 (1) which flew in August, 1939, powered by a Heinkel Hirth HeS 3B engine. In May, 1941, Whittle's engine, the Power Jets W.I, took to the air in the Gloster G.40 (2) Britain's first jet.

A month later, the Germans flew the world's first twin-jet, the Heinkel He 280V, and the tiny Messerschmitt Me 163B Rocket Fighter. The outstanding jet fighter of the war was the Messerschmitt Me 262 (3), which was way ahead in many respects. It had a top speed of 527 m.p.h., an ejection escape seat, a computer gun sight and air-to-air rockets. Luckily for the Allied daylight bombers, Hitler himself ordered its use as a *bomber*, which took away all its advantages until it was too late to affect the war in the air.

Britain flew the Gloster Meteor, which was in service in the last months of the war and, in 1949, came the brilliant English Electric Canberra (4) Bomber, which in its later versions is still in service after twenty-one years!

In 1947, the prototypes of a new breed of fighter flew in for the first time. They had swept wings which, as the Germans had discovered, allowed a great increase in maximum speed. The North American F-86 Sabre (5) and the Russian Mig-15 clashed over the battlefields of Korea in the early 1950's in the

first large-scale jet-against-jet combats. The Sabre gained mastery.

At the same time two superb pioneer civil jet designs were coming into airline service, the De Havilland Comet (6) and the Vickers Viscount (7). The Viscount was powered by Rolls-Royce Dart Turboprops, among the world's most reliable engines, which harnessed the power of the spinning jet turbine wheels to drive propellers very economically.

Into service in 1958, came the first of the big jets, the Boeing 707 (8), followed closely by the similar Douglas DC-8, continuing the "DC" series into the jet age. These brought long-range jet transport to the world, carrying 170+ passengers at cruising speeds of 500 m.p.h., flying at 30,000 feet with ranges of up to 4,000 miles.

In the military field, new designs were flying. The American Boeing B-47 and B-52 Stratofortress and the British "V-Bombers," like the Avro Vulcan (9), filled the "global bomber" role, flying high at 600+ m.p.h., carrying the nuclear deterrent weapons, now gone underwater or underground.

Fighters like the English Electric Lightning (10) were designed

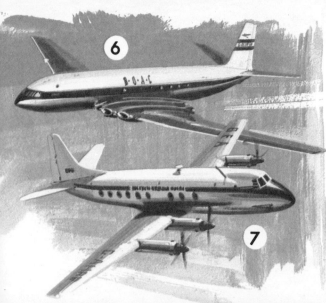

with the shape and power to slip with ease through the sound barrier and carry deadly air-to-air missiles deep into the sky at 1,500 m.p.h.

Among the fastest aircraft in the world today is the Lockheed YF-12A (11), a twin-jet monster, capable of well over 2,000 m.p.h. It is a masterpiece of science and engineering—as are the world-famous Bac-Sud Concorde Supersonic Airliner (12), and its first-to-fly Russian rival, the Tupolev Tu-144.

The two most breathtaking jet aircraft that aviation engineers have yet produced, are the North American XB-70A Valkyrie (13) and the Lockheed C-5A Galaxy (14).

The huge Delta-Winged XB-70A with six mighty engines was capable of reaching over 2,000 m.p.h., and of cruising at supersonic speeds for great distances. It proved its designer's advanced "compression lift" theory by gaining lift at high speeds from riding its own shock wave, as a motor boat rides its foaming bow wave.

The C-5A, at a weight of 365 tons and 246 feet in length is the world's largest aircraft. It can carry a 100,000 lbs. payload non-stop for 6,330 miles, and, although at present a military transport, could well revolutionise the air cargo industry in civil form.

One can only marvel at the brilliance of the designers of modern jet aircraft, such as have been described above. But one should not forget that it was Britain which took the lead in the development of pure jet transports. It is worth remembering that the de Havilland Comet 1, which flew for the first time on 27th July, 1949, was the first jet airliner in the world designed as such. Two weeks later came a Canadian project, which was largely British inspired. This was the Avro Canada C.102 Jetliner which had British engines but did not get beyond the prototype stage.

In May, 1952, BOAC brought the Comet into service but it was withdrawn after two years because a series of accidents had revealed defects. It returned, strengthened and enlarged in 1958 to fly the air routes of the world.

Although the Viscount and the Comet were important developments in aviation, they were only two of the many types developed after the war. A wide variety of designs were undertaken to meet varying requirements and the majority were highly successful. However, the supersonic era is now upon us with even more adventurous aeroplane designs in the offing.

VTOL

Although Leonardo da Vinci designed a form of helicopter, we must come down the centuries to 1923, to find the first successful rotating wing machine, the Cierva C.4. This was an *autogyro*—in which the rotor blades are not driven by the engine but free-wheel round as the craft is drawn forward by its conventional propeller. The first flying *Helicopter*—in which the engine drove the rotors round to provide lift *and* forward movement of the aircraft, was the Focke-Achgelis Fa 61 of 1936(1). The first truly successful helicopter was the VS-300 (2), designed and flown in 1942, by the giant among helicopter pioneers, Igor Sikorsky.

While helicopters progressed, many engineers were searching for ways of converting the enormous power available from jet engines and turboprops into systems for VTOL Aircraft which, once in flight, would have high-speeds. The Rolls-Royce "flying bedstead" (3) of 1953 was a test rig for examining methods of controlling jet thrust for VTOL.

In 1954, the American Turboprop Convair XFY-1 "Pogo," made successful complete flights with vertical take-off and landings, flying level at over 500 m.p.h. in between. In 1957, the Ryan X-13, (4), also American, did the same with jet power.

Scientists and engineers are now working on three main VTOL methods:—

TILT WINGS—as used in the Canadair CL-84 (5), in which the whole wing and engines are moved through 90°+ for vertical and highspeed horizontal flight.

VECTORED THRUST—as in the Hawker-Siddeley Harrier (6) in which the thrust of a powerful jet engine is directed through four swivelling exhaust nozzles to give VTOL and jet-fighter forward speeds, and

LIFT FANS—as in the Ryan XV-SA (7), in which the jet thrust is used in spin powerful fans in wings and fuselage to give downward thrust for VTOL. Once in flight, the fans are sealed over, and the jet engine thrusts the aircraft forward normally.

Jet VTOL may one day free us from aerial traffic jams and sprawling miles of concrete runways. Meanwhile, conventional helicopters get bigger and better every day. "Flying Cranes," like the mighty Russian MIL Mi-10 (8), carry enormous loads.

TO THE EDGE OF SPACE

As soon as man could fly, being man, he wanted to fly faster and higher—altitude and speed records have been consistently broken many times in the 67 years of flying, first simply to capture a record, then later to get answers to vital questions for future design work. An outstanding achievement came in June, 1937, when Flight-Lieutenant M. J. Adam, Royal Air Force, took a specially-built wooden aircraft, the Bristol 138 (1) to a height of 53,937 feet. To protect him, he wore a crude ancestor of today's space pressure-suits, with a huge can-shaped helmet.

After World War II, research flying began in earnest. The first of America's "X" (for experimental) series, the Bell X-1 (2) was designed to crash the sound barrier, at that time a truly frightening invisible wall in the sky, which hammered and buffeted any aircraft which approached it. Under average conditions, the speed of sound is 763 miles per hour at sea level, but this speed drops with altitude, and aircraft of 1945-1946 could approach it—with unpleasant results for the pilots, as the air compressed in front of them into a real barrier.

The X-1's designers were convinced that a 'plane the right shape, and with enough power, *could* go through the barrier. They modelled their fuselage shape on that of a ·50 calibre bullet, which they knew behaved well at 700 m.p.h. After a long series of tests at increasing speeds, a young Air Force Captain, Charles "Chuck" Yeager, now a legendary test pilot of his own time, made the attempt on October 14th, 1947.

Dropped at 30,000 feet from a B-29 "mother" 'plane, he cut the rocket engine and flew into the unknown. First came the buffeting, then instability—then calm and silence—he was through the sound barrier and there were no "demons" on the other side.

Speeds and heights increased, each series of flights bringing back more knowledge. The ultimate piloted "X" aircraft was the North American X-15 (3), in which several pilots gained astronaut's wings! It reached speeds of over 4,000 m.p.h. and heights over 67 miles on rocket power and needed two sets of

controls—conventional ailerons, elevators and rudders for flight in atmosphere, and gas jets like a spacecraft for manoeuvring in the vacuum of the edge of space. It was the first aircraft to undergo the blistering heat of re-entry into the earth's atmosphere.

The "space shuttles" of the future may be wingless lifting bodies, like the experimental M2F2 (4), carrying men to and from space stations, flying in the atmosphere by the airflow over their bulbous shapes...

On April 12, 1961, the late Yuri Gagarin opened the new age of flight. In his tiny Vostok I Spacecraft (5) he was shot into orbit around the Earth, 200 miles up, and travelling at 17,000 miles per hour.

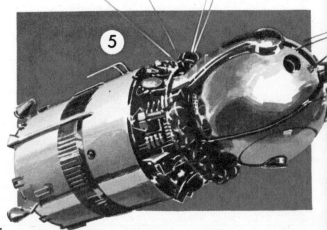

THE BRIDGE BUILDERS

Bridges make you gasp (like the Forth Road Bridge, facing page) or charm you with their beauty. But we could not do without them, as you will discover.

ONCE upon a time, a long while ago, if you wanted to cross a river, you either had to swim across or pick your way over on precarious stepping stones. Then, man had an idea. He saw that when a tree, blown down in a storm, fell with its branches on one bank and its roots on the other, a convenient way of getting from one side of the river to the far side had been presented to him.

Bridges had been born. This kind was called a beam bridge and it was later built of shaped logs and planks. These bridges, however, were limited in size by the strength of the wood. But with today's steel girders, much greater spans are possible and many simple bridges are of this type.

When the *arch* was discovered, not only could larger spans be used, but much heavier loads could be carried. Masonry arches, however, became very heavy in long spans, and modern arch bridges, of which the Sydney Harbour Bridge in Australia is a good example, are made of steel.

Another kind of bridge is the *cantilever* which is rather like a man with both arms outstretched and a weight in each hand. The Forth Railway Bridge in Scotland is of this type. Between each pair of cantilevers may be placed a light girder bridge. The cantilever gives a long span and good headroom.

Then there is the *suspension* bridge in which the roadway is suspended from a continuous cable.

Above: old London Bridge in the 18th century. Below: work in progress on the construction of the latest London Bridge which will carry today's heavier traffic into the capital.

This, held by high towers on each bank, runs completely across a river without any intermediate support. With modern materials, very long spans are possible, giving good headroom for shipping.

Transporter Bridge

Another kind of bridge is the *transporter*, which is supported at both ends on pylons, with a travelling cradle on which passengers and vehicles are carried across.

All of these various kinds of bridges have served the very valuable function of making it easier for people to get about. It is almost impossible to travel anywhere by road or rail without crossing a bridge of some kind and big cities like London would be split were not the River Thames spanned by many bridges.

The most famous of these is London Bridge, which everybody has heard of, even if only through the old nursery rhyme. Many bridges had been built on its site until the first all-stone bridge was finished in 1209. On it were a number of timber houses and it seemed more of a continual street than a bridge.

This stood until the 18th century when it was replaced by a new one designed by John Rennie. Apart from widening in 1902–4, this

L.

D

and are 505 ft. high and 78 ft. apart.

Across the bridge run two cycle tracks, two footpaths and two roadways $1\frac{1}{2}$ miles long.

We would have to fly to the other side of the globe to see the largest, although not the longest, steel arch bridge. This is the Sydney Harbour Bridge in New South Wales, Australia. Its main arch span is 1,650 ft. long and its deck—that part of the bridge which makes up the roadways—is 160 ft. wide.

The total length of the bridge, counting both approaches, is two-and-three-quarter miles. Sydney's northern suburbs and the city itself are linked by this bridge's two railway tracks, eight lanes of motorway, a lane for cycles and a footpath, all of which are 172 ft. above the waters of the harbour.

The Australians miss the honour of having the longest steel arch bridge by only 25 inches, being beaten by Bayonne Bridge, New Jersey, U.S.A.

To see the beginning of such bridges as these we have to go back to the end of the 18th century when the Machine Age had begun. Men with the minds of giants were needed to build canals, roads, bridges and docks on a scale never attempted

bridge stayed until it was recently decided to replace it with a modern bridge.

For many years, the most famous bridge in Scotland was the railway bridge over the Firth of Forth, where the River Forth widens as it nears the sea. This bridge was begun in 1882 and opened in 1890.

But in 1964, another bridge over the Firth of Forth was completed. This was the Forth Road Bridge which is a suspension bridge with a central span of 3,300 ft. and spans at each side of 1,340 ft.

The main towers are made up of two hollow steel legs joined together at intervals by braces. They stand on concrete cylinders sunk about 100 ft. below ground level

Sydney Harbour bridge in Australia is the largest but not the longest steel arch bridge ever built.

before. They had to span wide valleys, link up rivers and tame mountain and moorland.

A man who fitted into this setting was Thomas Telford, a great civil engineer, who was one of the first to use iron for bridge building. He found that it gave a flatter arch and put less weight on the foundations than stone.

Telford built at least a hundred bridges and would have gone down into history even if he had not built his graceful bridge across the Menai Strait in Wales. Constructed on the suspension principle, 1,170 ft. long and a hundred feet above the high water mark, it was the largest of its kind.

In the midst of success, Telford could be humble. When his friends sought him out to congratulate him as the first London to Holyhead mail coach drove across the Menai Bridge, they found him on his knees, filled with gratitude and relief.

Telford believed that a good appearance and strength could be combined in engineering, but in none of his bridges did he attempt to equal the beauty of the Bridge of Sighs in Venice, Italy. This was built between 1595 and 1605 by the Italian architect, Contino.

It connects the palace of the doge (or chief magistrate) of Venice and the dungeons of the state prison. At one time, when prisoners had been condemned by the magistrate,

they were led across the Bridge of Sighs to the prison. Since their view of the city through the windows of the bridge was their last for many years or—in the case of those condemned to death—for ever, their sighs must have been loud indeed.

Today, however, bridges do not usually provoke sighs, but are accepted as a normal feature of modern transport.

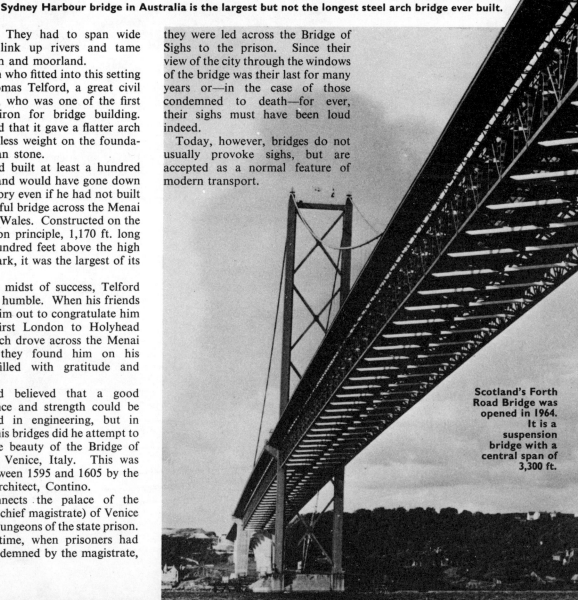

Scotland's Forth Road Bridge was opened in 1964. It is a suspension bridge with a central span of 3,300 ft.

IT'S MAGIC

How the magician uses science to baffle his audience.

THE art of conjuring, or "white magic," is very old indeed. In fact, no one knows when it all started; probably it began with the rise of Man himself.

Perhaps one of the earliest of all conjuring tricks is the "cups and balls" illusion. This trick is still performed today, and requires three small cups about three inches deep, and a number of balls. There are usually more than four balls, sometimes fifteen or sixteen, in three different sizes.

Three cups are placed mouth downwards on to a table. One of the cups is lifted to show that there is nothing beneath it, and is then replaced on the table. A ball is picked up with the right hand and placed into the left hand, while the magician orders it to go under the inverted cup. When the magician opens his hand, the ball appears to have gone, and on his lifting the cup, the ball is found underneath. This routine is followed so that all the cups have a ball beneath, which then reappear and disappear

apparently at the magician's will.

The trick needs no complicated apparatus, and depends entirely upon the skill of the performer in "palming" the ball. This is called sleight-of-hand, in which magicians are able to hold objects in their hands in such a way as to convince their audiences that there is nothing there.

Stage magic made little progress during the Middle Ages; indeed, conjurers were usually classed amongst rogues and vagabonds, and were not regarded as fit company for respectable people. They were suspected of dark deeds, of dealing in witchcraft, and perhaps of consorting with the Devil.

SAWING A HORSE IN HALF

As soon as the horse goes inside, a dummy head and a dummy tail appear from each end of the box. These dummies are pulled into position by wires which are invisible to the audience, and continue right up into the flies. During the performance, an assistant moves the wires so that the head and tail appear alive. The horse actually goes inside a container, which is then lowered to the floor of the box by means of cables leading to a winch below the stage. This means that when the saw goes through the centre of the box, it stops at the top of the lowered container.

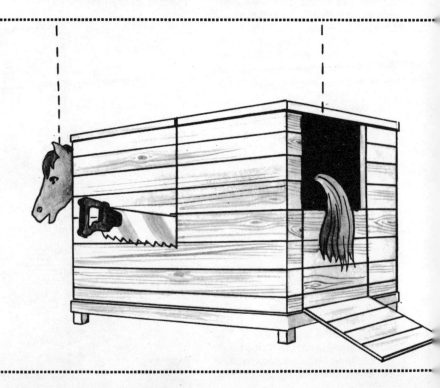

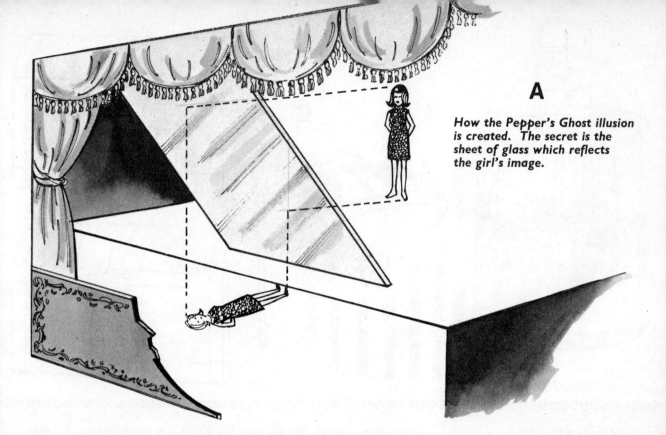

A

How the Pepper's Ghost illusion is created. The secret is the sheet of glass which reflects the girl's image.

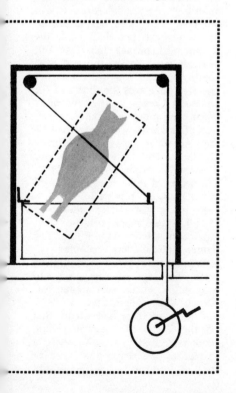

As early as Chaucer's time, people were astonished by miraculous images produced by a kind of primitive magic lantern, but it was not until about the 18th century that the art of conjuring began to be taken at all seriously.

Magicians then started to use mechanical devices to produce illusions, and instead of being mere wandering performers, set up stages, and wore spectacular clothing to heighten the effect of mystery. A table, covered with a cloth, became a standard piece of stage property.

The cloth was draped over the table so that it practically covered the sides, in a somewhat clumsy attempt to hide the presence of a small assistant beneath. Despite these limitations, conjurers were able to produce some quite convincing effects.

Towards the end of the 18th century, an Italian conjuror, Giuseppe Pinetti, invented a table which required no assistant underneath. Instead, his table was fixed to the floor, and a series of wires ran from the top of the table, through the legs, to a cabinet behind the scenes, from which an assistant could, by remote control, produce remarkable effects on the conjuror's table.

The true father of modern magic was Jean-Eugene Robert-Houdin, who was born in 1805 in France. He was at first a watchmaker, but became interested in stage magic later. When he was forty, he opened his own theatre of magic in Paris, and from there his ame spread all over the world.

Robert-Houdin was outstanding, because of his use of scientific inventions to produce magical effects. He was the inventor of a "light and heavy box," which mystified everyone. The curtains of the stage parted to reveal a chest or box, and members of the audience were invited on to the stage to lift the chest. This could be done quite easily, but Robert-Houdin would then ask the volunteers to try again, when the feat proved to be impossible. Somehow, the chest had become tremendously heavy.

The explanation was quite simple. Robert-Houdin had placed an electromagnet beneath the surface of the stage, which he could switch on at will. This electromagnet acted on the bottom of the chest, which was made of steel.

One of the first modern English masters of magic was John Nevil Maskelyne, who, like Robert-Houdin, began his career as a watchmaker. He was born in 1839, and after becoming interested in stage magic, teamed up with his friend, George Cooke, to perform some amazing illusions in the famous Egyptian Hall in London.

One of the most famous illusions of the nineteenth century was known as Pepper's Ghost, after its inventor, Professor Pepper. Audiences watching the stage saw queer ghostly figures of people appearing and disappearing in mid-air. The illusion

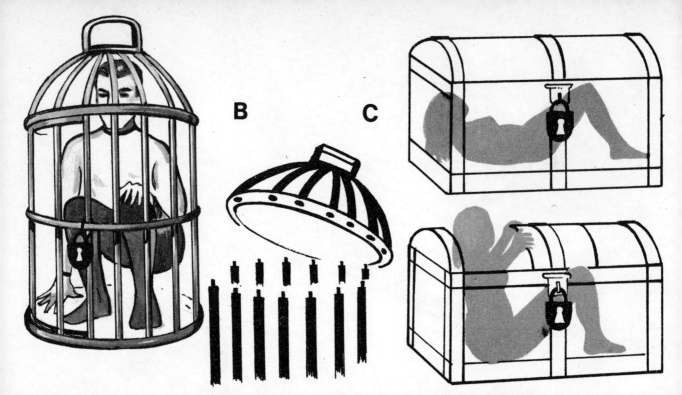

B

C

was produced by making a gap in the floor of the stage just behind the base of the proscenium, invisible to the audience.

A large sheet of glass was then placed so that the top was concealed behind the curtains at the top of the proscenium arch, and the bottom fixed at the back of the stage, so that the glass inclined at an angle of 45 degrees towards the audience.

A girl lay down in the gap, and a light was shone on to her, which reflected an image into the glass. The effect produced to the audience was of a figure apparently floating in the middle of the stage. By dimming the lights, the figure could be made to disappear. (See picture A.)

A similar trick was the "Man without a Body." A cabinet, about six feet square, and about eight feet high was shown on the stage with black curtains drawn all round it. The curtain was then drawn back, and the audience saw a man's head apparently floating inside without a body attached. Even more amazing was the fact that the man laughed and joked with them.

The illusion was produced by placing a mirror inside the cabinet, large enough to reach to the back of the cabinet at the top, and to the front edge of the platform at the bottom. This was inclined at an angle of 45 degrees away from the audience, and had a hole cut in the

centre large enough for a man's head to go through. A man then stood behind the mirror, and put his head through the hole, wearing a ruff to hide the opening. The effect was quite remarkable, since the audience saw an apparently empty cabinet with a head floating in it.

"Sawing" a Woman

One of the most famous tricks in the world is the well-known illusion "Sawing a Woman in Half." This was invented in 1921 by Percy Thomas Tibbles, an Englishman, who began life as a journalist, and then became a magician under the name of P. T. Selbit. Several variations of the illusion were performed at the time, and the subject is still one of the most popular among audiences.

Resting upon two trestles or chairs is a long box, and a girl assistant gets into the box and lies full length in it. Her head protrudes from one end, and her feet protrude from the other end. Sometimes a member of the audience is invited up on to the stage to hold her ankles. Usually, the magician passes a rope through the box to fasten the girl's wrists and ankles.

The magician then takes a saw, and apparently saws right through the middle of the box. Two metal slides are then placed on either side of the saw cut, and the two halves

of the box are pulled apart. The halves are then pushed together again, the slides are taken out, and the girl comes out of the box unharmed.

There are several ways of doing this trick, but usually there are two girls involved. The bottom of the box appears to be flat and level, but actually, this surface is tapered in such a way as to allow room for the other girl to hide beneath the main box. As the first girl gets into the box, she draws her knees up tightly towards her chin, while the second girl pushes her legs out through the holes at the end, and bends her head forward on to her knees. This leaves the centre of the box empty, ready for the saw cut.

An even more remarkable variation of this trick was performed by the American magician Horace Goldin. This was the fabulous illusion "Sawing a Horse in Half." The secret of this trick is explained in the pictures on page 36.

Probably the most well-known of all stage magicians was the American, Harry Houdini. The name Houdini has become synonymous with brilliant conjuring since the days of the successful originator, whose real name was Ehrich Weisz. He was born in 1874, and took his stage name from that of the earlier French magician, Robert-Houdin. Houdini was a master of the sensational type of trick, but he was also a past master

D

The "floating girl" trick never fails to mystify audiences, but the girl is supported by a powerful steel rod, as these pictures show, and not by any magical influence, as the conjuror implies.

of escapology, which is concerned with the ability to escape from apparently impossible conditions.

One of the more well-known "escape" illusions is the "Man in the Cage." A heavy iron cage is shown to the audience, and members are invited up on to the stage to examine it. The cage stands on a small platform, and the performer gets in. The door is closed, and a padlock is placed through the hasp and locked. A screen is then put in front of the cage, and after a short pause, this is removed, to show the performer out of the cage and free.

Once again, the members of the audience are asked to examine the cage, when they find that the door is still locked and intact.

As with all good illusions, the secret is simple. The top of the cage lifts off, but it is so heavy that no one examining it could possibly lift it off; and in fact, no one ever tries. The performer has to be a very strong man, able to lift the top of the cage with his shoulders. Having lifted the top clear, he manoeuvres it back on to the cage, and lets it drop back gently into position. Usually the orchestra plays loudly during this event, to hide any possible noise made in lifting the top of the cage. (See picture B.)

Another "escape" trick is the "Girl in the Trunk." A cabin trunk is brought on to the stage, which has glass panels all round, with curved glass panels in the top. Two

members of the audience are invited to examine the trunk and then a girl inside it. The lid is closed and secured with a padlock, the key is given to a member of the audience, and a screen is placed in front of the trunk. A minute or so later, the screen is removed, and the girl is seen quite free.

What happens is that one of the curved glass panels on the top is slid along to make a small opening. The girl can free this panel from the inside by turning a screw. She has a duplicate key for the padlock, which she unlocks, opens the lid and steps out. She restores the trunk to its locked condition, and reappears to the audience. (See picture C.)

Floating girl

"The Floating Girl" is a very spectacular trick, in which a girl appears to be suspended in mid-air. The magician places a board between two chairs, a girl lies full length upon it, and is covered with a cloth. The chairs are removed one by one, and the girl is left apparently floating without support. The magician passes hoops around her, and this seems to prove that nothing is holding her up.

In fact, the girl is being supported by strong steel bars bent into a pattern which will allow the hoop to pass with the minimum of hindrance. (See picture D.)

Sometimes, the trick is done with the metal support passing through one of the chairs, in which case, only one of them can, of course, be removed.

One trick which is always very popular with conjurers and audiences is the "Magic Bottle." The performer produces what looks like an ordinary wine bottle, and pours a little red wine from it into a glass. Then, from the same bottle, he pours some white wine, followed by milk, water, and lemonade. The bottle seems to go on pouring for ever, and the magician can apparently change the liquid at will.

The secret is in the magician's bottle, which is quite a complicated piece of apparatus. It is made of an opaque substance, such as plastic, and is divided into five compartments, each containing a different liquid. Tubes lead to the top of the bottle, and to the air-holes near the neck. When the magician covers an air-hole, no liquid will flow, so that he can choose any particular liquid simply by uncovering its air-hole.

Some of the glasses into which the magician pours water are prepared previously with a few drops of a chemical, which changes colour when water is mixed with it. In this way, the illusion becomes even more spectacular, ending with dozens of small glasses all containing different coloured liquids.

WATER, WATER, EVERYWHERE

STAND in a highland valley and look around you. The mountain peaks above you were shaped by water, and water is still at work sculpting and levelling them. Perhaps the last remnant of the winter's snow gives their crests a sharp outline against clouds that promise rain. The valley itself was carved by water, and in the pools of the stream curling through the rocks are fish whose life depends on air dissolved in the water. The trees by the stream consist mainly of water, and through its roots and leaves each one may transfer fifty gallons of water from soil to atmosphere during a summer's day. The stream carries a load of solid particles, the remains of broken-down rock, which will be deposited to form fertile alluvial flats when the rushing mountain torrent becomes a sedate lowland river; and it also carries a further load of dissolved material which it will carry down to the sea. And from life in the sea you, the watcher, slowly evolved over millions of years; and you too are mostly water, and your body fluids contain salts like those in sea-water.

The behaviour of water in large quantities depends, of course, on the properties of the individual particles, or molecules. Although the molecule of water appears to be very simple, containing only two atoms of hydrogen and one of oxygen (which is why it is given the well-known formula H_2O), this combination of atoms results in many unusual properties. Among these are a melting point and boiling point very much higher than is normal for such a small molecule, and a remarkable ability to dissolve electrolytes (i.e. those substances which when molten or dissolved in water can both conduct an electric current and be broken up by it).

But electrolytes, such as common salt (sodium chloride), are not the only kind of compound that can dissolve in water. Sugar, for instance, is not an electrolyte, but many people make use of its solubility in water every time they drink a cup of tea. Water dissolves at least small amounts of practically everything, and so it is extremely difficult to get it pure. Even Pyrex glass dissolves to some extent, so that

Left: far beneath the earth's surface, water carves out marvellous caves—and provides an extra hazard to exploring potholes. Above: the valley carries the rushing streams cascading from the mountains, themselves made by water. Right: Power from water. The huge man-made Lake Mead in the U.S.A. is formed by the Hoover Dam. Below: vast quantities of "solid water" are contained in the icebergs which drift in the Polar regions of the earth.

distillation in glass apparatus in not enough to make water really pure.

When rain falls on the land, soluble materials in the soil dissolve and are carried by the rivers into the sea. Several of these dissolved minerals cause trouble by forming scale and "fur" in, for example, boilers, hot water pipes, and even electric kettles. Water which forms these deposits easily (and, incidentally, also forms a scum with soap) is called "hard". water. The "hardest" water is found in areas where the local rock is mainly chalk or limestone. Most of the "hardness" is caused by rainwater slowly dissolving some of the chalk or limestone; the very weak solution formed in this way is broken up when the water is heated, and the solid is re-deposited in the boiler, pipe, or kettle. In the course of many centuries this slow dissolving of limestone can produce marvellous caves, with their stalactites and stalagmites, and such majestic pieces of scenery as Malham Cove in Yorkshire and the Cheddar Gorge near Bristol.

Most of the dissolved substances, however, are carried by the rivers down to the sea; and there they must stay, while heat from the Sun evaporates water from the surface of the sea to form water vapour in the atmosphere. This vapour eventually condenses and falls as rain, and the cycle is repeated: rain, river, sea, water vapour, rain. So more and more dissolved substances are carried down to the sea, where they stay. The sea contains about 3·6 per cent by weight of dissolved solids, and the amount is very slowly increasing. Each cubic mile of sea water contains, for example, about 130 million tons of salt and 0·028 tons of gold, besides vast amounts of other materials. An extreme example of the process is found at the lowest point of the Earth's surface, where the River Jordan flows into the Dead Sea. In the intense heat of the desert, water is continuously removed from the Dead Sea by evaporation; the salts have nowhere else to go, and so the Dead Sea contains more than 20 per cent by weight of dissolved solids. The density of the water is so great that a bather cannot get his body deep enough in the water to swim properly.

Water is easily the most common chemical compound found on Earth: and without it life cannot go on. Usually, more than 80 per cent of the weight of a living cell consists of water, for water is the only solvent in which the many reactions necessary for life can take place. Also, water is vital for the support of plant tissues—without water, plants simply collapse —and it is an essential raw material for the all-important process of photosynthesis. In this reaction, carbon dioxide from the atmosphere and water from the soil are joined together in green plants by the action of sunlight. The plants are green because they contain the green pigment chlorophyll, and without chlorophyll the reaction cannot happen. Photosynthesis provides the sugar-type compounds which form the basis for plant growth. Remember, plants provide food for animals; so whether or not liquid water is available decides where plants can grow and animals live on the land surface of the Earth. Irrigation—that is the bringing of water to places where not enough rain falls for plants to grow well—is an activity that concerns very many people. A shortage of fresh water for drinking and for irrigation is one of the likely consequences of the rapid increase in the human population of the world.

So dense is the solid content of the Dead Sea that the swimmer finds it almost impossible to swim.

Water, like any other material, contracts on cooling; but, unlike almost everything else, it expands again on cooling from liquid (at 4°C) to solid (at 0°C). So, ice has a lower density than water, and floats. If this were not so, the Arctic and Antarctic seas might well freeze solid: ice, when formed, would sink to the bottom instead of forming a good insulating layer on the surface. Another result of this expansion on freezing is increased employment for plumbers in a hard winter: the water in a pipe freezes, expands, and splits the pipe. The leak only becomes obvious when a thaw arrives. The same thing can happen to the cylinder block of a car engine unless the radiator water contains anti-freeze. It is this expansion of water on freezing which is even now changing the shape of our hills and mountains: the basic outlines were determined by the sheer abrasive force of the slow-moving glaciers of the Ice Ages, but much of the erosion since has been due to splintering and chipping of the rock by the great expansion forces of freezing water.

Rapidly flowing water has been used as a source of power almost since civilisation began. The old-time water-wheel has now been replaced by the turbines of modern hydro-electric power stations, where the energy of falling water is changed into electrical energy. Water, in the form of steam, is the working fluid in other power stations; heat energy, obtained either by burning coal or oil or from a controlled nuclear reaction, is used to convert water into super-heated steam which drives the turbines that generate the electricity. Compared with most other substances, it takes a lot of energy to raise the temperature of water and convert it into a gas, and a corresponding amount of energy is given up when the steam condenses. One consequence of this resistance to temperature increase is that the ocean is slow to warm up in the summer and slow to cool down in the winter; this has very great effects on our climate, including the well-known fact that coastal districts have cooler summers and warmer winters than inland districts.

To the farmer, lack of water is as great a hazard as having too much. A period of drought can cause untold suffering.

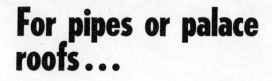

For pipes or palace roofs...

Copper is King!

Copper is one of the most important metals in the world. Egyptian craftsmen used it 7,000 years ago...and today it is vital in space travel research. Electrical and electronic apparatus would not function without it, because wire as fine as that shown on this page can be drawn from it. In contrast, it can be made into sheets vast enough to cover a large roof. Because the Romans obtained their copper from Cyprus, they called it cuprum, from which our word is derived. Its family of alloys includes brass, bronze, nickel silver and aluminium bronze.

PLEASE TURN TO NEXT PAGE

This greatly enlarged drawing of the eye of a needle shows the comparative thickness of a cotton thread with a fine copper wire.

Top: Miners in a Canadian copper mine dig out the ore. (Right): These cylindrical machines are ball mills through which the copper ore passes in its first stage of concentration.

More processes follow that at the top of this page, but finally the copper is cast into bars ready to be sent to huge mills to be rolled into rods.

*C*OPPER is the most exciting and romantic of all the common metals. Not only was it the first metal used extensively by man, but it is still among the most widely used and important metals in industry.

Historians believe that this metal was first worked by the early Egyptians in the neighbourhood of the rivers Tigris and Euphrates as early as 6000 B.C. Copper objects have been found in Egypt and in Iraq in earth below the level of the Great Flood of about 4000 B.C.

These early articles were almost certainly made by hammering out pieces of native copper, but before long it was found possible to melt this natural metal and to cast it into shape. The method of smelting an ore, such as malachite, to extract the metal followed, probably about 3000 B.C.

In Britain, the copper industry was greatly enlarged by the Romans, but it declined during the Dark Ages from the 5th century, A.D. until the beginning of the 17th, when the modern industry of copper smelting from the ore was founded in Great Britain and rapidly advanced.

Among the early products were plate and wire, followed by brass foundry work such as toys and buckles.

Later, it became firmly established

in the electrical industry as a good conductor, and it is still popular in architecture and art, as well as having a wide use for metal alloys used in industry.

Nowadays, about five million tons of copper are produced each year from two principal ores: sulphide ores and oxidised ores.

Most of this is produced in America (about 30 per cent), followed by Zambia and Russia (15 per cent each), Chile (12 per cent), Canada (11 per cent) and the Congo (8 per cent). The consumption throughout the world works out at about 2 lb. a person. However, Britain and Sweden are the largest users with about 25 lb. per head of population.

Ores considerably in excess of the copper produced are smelted, or treated by another method to extract the copper, for the ores often contain such other substances as iron, zinc, silver, mercury, antimony, arsenic and manganese. Only about 4 per cent of the ore is copper.

The method by which this is extracted depends upon the nature of the ore and the other metals or minerals it contains.

In what is called the "wet" method the ore is crushed and the waste materials washed away by water.

Smelting—a way of extracting metal from the ore by melting it in a furnace—gets rid of any remaining impurities and separates the refined metal from these. This metal is cast

The machine above rolls copper into strips into the thicknesses manufacturers need. It is sent to them in coils or rolls. Below: brass and copper are made into more than 5,000 different shapes in the rod mill shown here.

45

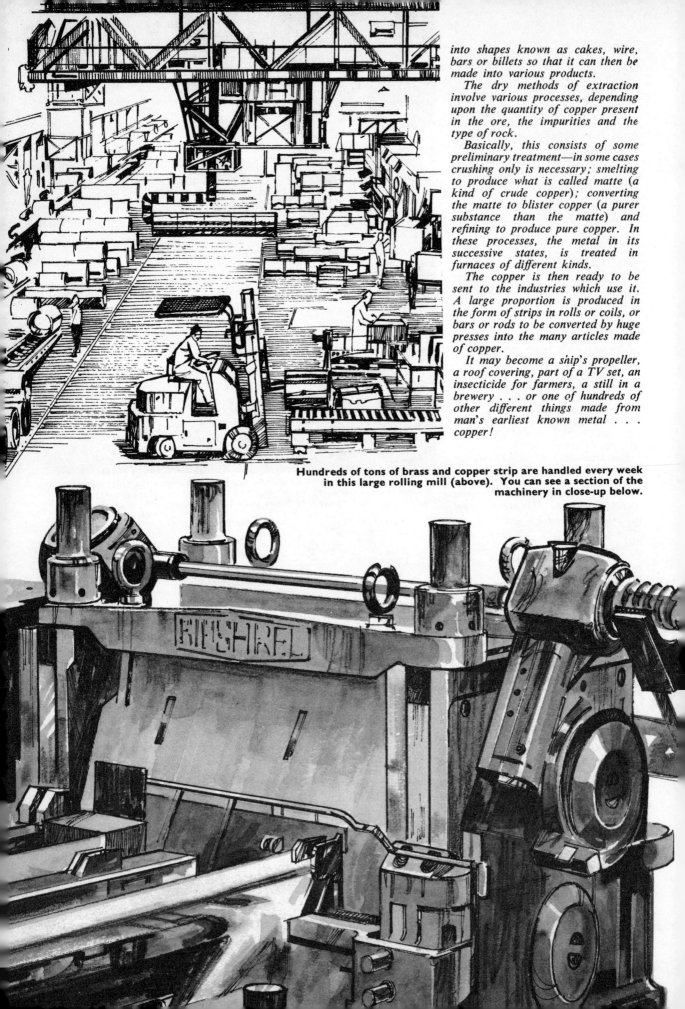

into shapes known as cakes, wire, bars or billets so that it can then be made into various products.

The dry methods of extraction involve various processes, depending upon the quantity of copper present in the ore, the impurities and the type of rock.

Basically, this consists of some preliminary treatment—in some cases crushing only is necessary; smelting to produce what is called matte (a kind of crude copper); converting the matte to blister copper (a purer substance than the matte) and refining to produce pure copper. In these processes, the metal in its successive states, is treated in furnaces of different kinds.

The copper is then ready to be sent to the industries which use it. A large proportion is produced in the form of strips in rolls or coils, or bars or rods to be converted by huge presses into the many articles made of copper.

It may become a ship's propeller, a roof covering, part of a TV set, an insecticide for farmers, a still in a brewery . . . or one of hundreds of other different things made from man's earliest known metal . . . copper!

Hundreds of tons of brass and copper strip are handled every week in this large rolling mill (above). You can see a section of the machinery in close-up below.

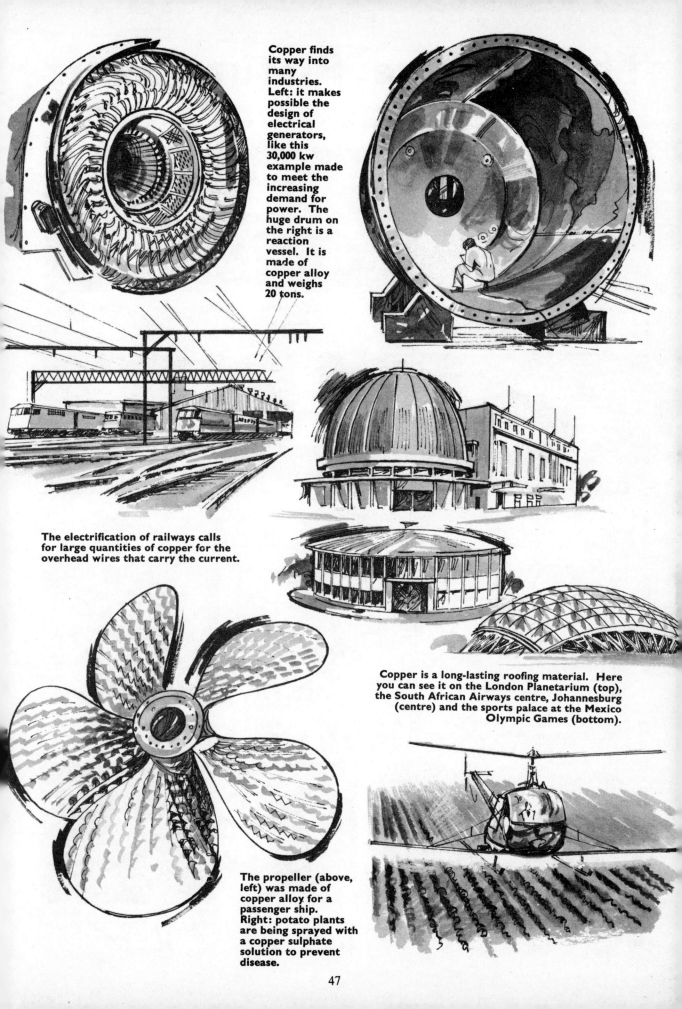

Copper finds its way into many industries. Left: it makes possible the design of electrical generators, like this 30,000 kw example made to meet the increasing demand for power. The huge drum on the right is a reaction vessel. It is made of copper alloy and weighs 20 tons.

The electrification of railways calls for large quantities of copper for the overhead wires that carry the current.

Copper is a long-lasting roofing material. Here you can see it on the London Planetarium (top), the South African Airways centre, Johannesburg (centre) and the sports palace at the Mexico Olympic Games (bottom).

The propeller (above, left) was made of copper alloy for a passenger ship. Right: potato plants are being sprayed with a copper sulphate solution to prevent disease.

47

The Science Picture Quiz

Now that you have read the Science section, see if you can answer this simple picture-quiz.
The answers are at the back of the book.

1 In what year did man first become airborne in a balloon? What kept it in the air?

2 Who were the world's first airmen in 1903 and what was their record flight then?

3 Which country had a jet plane in 1939? When did Britain's first jet aircraft fly?

4 Do you know the name of the bridge being rebuilt here? It is on a site made famous in rhyme.

5 Which is the largest, but not the longest, steel arch bridge in the world? Do you know where it is?

6 What bridge did Thomas Telford build in Wales?

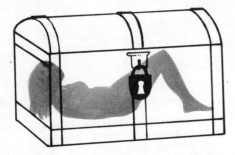

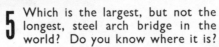

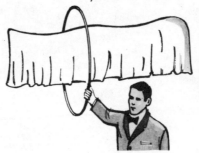

7 For a person locked inside this chest, escape is easier than it seems. How is it achieved?

8 A woman covered with a cloth "floats" in the air. What keeps her "airborne"?

9 Can you say what causes valleys to be formed?

10 This inland sea is at a lower level than any other water. What is its name and what gives it its unusual buoyancy?

11 These men are mining copper. What did the Romans call this metal?

12 How much copper is produced each year?

ROTORUA – THERMAL WONDERLAND

There are often Maori guides at the spouting geysers and steaming pools.

Spectators can get really close to the geysers, and normally there is no danger.

ONE area of the world where nature can be seen at her most exciting is New Zealand's thermal region. Situated on the North Island, with Rotorua at its centre, it seems almost aggressive in the way it hurls its splendour at the visitor.

Geysers rumble and roar as plumes of boiling water and steam hiss skywards. Mud pools gurgle and plop in captivating displays of bizarre shapes and colours. White vapour wisps and billows from steam fumaroles, and crystal clear hot springs bubble and spurt as they toss cascades of scalding water down terraced formations.

One of the most interesting geysers in Rotorua is to be found at the Whakarewarewa Thermal Reserve, a Maori settlement where villagers have become so used to the natural phenomena that they calmly wash clothes and cook food in the boiling pools. The sightseer, however, is more likely to stand back and gaze in awe at the Prince of Wales Feathers, which tosses up plumes of scalding water as an opener to the display by the famous Pohutu geyser. Forever rumbling and bubbling, it suddenly hurls a fountain of boiling water to a height of about 70 ft.

Within about 20 miles of the city of Rotorua even more splendid and startling geysers are to be found. And the most amazing sights of all, perhaps, are to be found in the Waimangu Thermal Reserve, a valley of such overwhelming spectacles that more susceptible people believe it to be a specially preserved slice of the days of creation.

Here are to be found Lakes Rotomahana, Tarawera, and the Blue and Green Lakes; and here, too, lies the buried village of Te Wairoa, which was destroyed when Mount Tarawera erupted in 1886.

Other attractions in the valley are the largest near-boiling lake in the world, the Echo and Inferno Craters, formed in the Tarawera eruption, Ruamoko's Throat, a bright blue volcanic lake that rises and falls fifty feet, and the Dragon's Fangs, a jagged outcrop on the steaming cliffs.

Six miles north of Taupo, New Zealand's largest lake, is Wairakei, and here man has contributed an attraction of his own. A number of bores were drilled deep beneath the surface, and the superheated steam there was harnessed to drive the turbines of a large electric powerhouse.

The thermal heart of New Zealand can be spectacular—it is also dangerous!

Nobody can safely say just how violent the area can be. Attention is often drawn to Mount Ruapehu which was considered extinct until Christmas Eve in 1953. Then its crater suddenly surged its banks, hurled down masses of boiling water and caused the deaths of 151 people on the Wellington-Auckland express.

Lake Rotomahana (above) is one of the most spectacular of the lakes.

In some parts of the thermal region, visitors can bathe in the lakes.

51

TIMBER!

"TIMBER!" The lumberjack's warning cry as another giant pine tree topples lazily sideways and crashes with a mighty thud to the ground, a further victim to the axes and power saws that feed Canada's great timber industry. Trees cover nearly one million square miles of Canada, and the strong, silent sentinels of the snow form one of the country's richest treasures. The forests of Canada extend in an unbroken belt, 600 to 1,300 miles wide, which stretches from the Atlantic to the Pacific. Apart from Siberia, Canada is the biggest timber country in the world.

But, despite its great wealth of trees, terrible fires started mainly through carelessness and the relentless felling of the woodsman's axe, are diminishing Canada's forests faster than the conservancy experts can replace the trees. It takes over half a century for some trees to

grow sufficiently to be of commercial value. Conifers grow quickly, are cut easily, and can be swiftly transported, but because there are so many uses to which they can be put, they are being rapidly depleted.

The fir forest of the Rocky Mountains and Pacific Coast, the northern coniferous forests stretching from north of the Great Lakes to Labrador, and the hardwood deciduous forest extending through Southern Ontario and Quebec to New Brunswick and the Atlantic coast, are the main tree areas. The timbers used for commerce are classified as 'softwoods,' which come from coniferous trees and 'hardwoods' which are derived from deciduous and evergreen trees.

Hundreds of different trees are found in Canada. There are the conifers, mainly western hemlock, red cedar and Sitka spruce, Douglas fir, Alpine fir, Engelmann spruce

and mountain hemlock. There are the deciduous trees such as basswood, maple, red maple, black ash, white ash, white elm, the list is endless. Building takes up a large supply of wood, then wood is needed in large quantities to turn into wood pulp for the manufacture of newsprint for America. Canada's forests could supply many of the countries who need pulp, paper and lumber for decades. The manufacture of pulp and paper has been Canada's leading industry for many years. Her many streams carry logs to the mills and also supply the hydro-electric power which drives them.

Canada as a country is rich in natural resources and one of the greatest of these is her timber. To keep it the lumberjacks have to practise selective logging, taking only mature trees. If her forests are well looked after there will always be plenty of trees for the world.

SHOWING THE FLAG

MOST people would be surprised to know that the American flag, the Stars and Stripes, has been depicted on a British stamp. The same stamp also showed the French blue, white and red tricolour, Denmark's ancient flag, the Dannebrog, and the flags of nine other nations. All the countries belong to the North Atlantic Treaty Organisation and the stamp was issued in 1969 to mark the Organisation's twentieth anniversary. Because it was a 1s. 6d. value intended mainly for use on airmail letters going overseas, the stamp was seldom seen on mail in Britain.

Nobody knows for certain when flags were first used. The Roman legions carried standards bearing a portrait of their Emperor and when a Roman general was preparing for battle, a red flag was hoisted outside his tent.

During the Middle Ages, when soldiers wore heavy armour, a flag, like a coat-of-arms, was a useful means of distinguishing friend from foe. For centuries their country's flag was a rallying-point for soldiers in battle and for the enemy to capture it was a dire disgrace. The Stars and Stripes was depicted on an American stamp as long ago as 1869 but only in recent years, since the large-scale production of multi-coloured stamps became possible, have flags come into their own. Now they may be seen in full colour on many attractive issues.

International events such as football championships, political conferences and the Olympic Games are frequently the occasions for stamps showing flags of the countries taking part. When nations achieve their independence, one of their first tasks is to choose a

8c

NEW ZEALAND

POSTES · POSTAGE 5

CANADA 1867 1967

COMMEMORATING TEN OLYMPIC YEARS

MELBOURNE 1956

4B POSTAGE

THE MUTAWAKELITE KINGDOM OF YEMEN

١٩٤٨

LONDON 1948

POSTAGE **4B**

COMMEMORATING TEN OLYMPIC YEARS

THE MUTAWAKELITE KINGDOM OF YEMEN

4¢ POSTAGE

LONG MAY IT WAVE

UNITED STATES OF AMERICA

3d

JAMAICA

national flag. The sun rising above a blue sea was an appropriate device for the West Indian island of Antigua when it became independent in 1967 and when Canada decided to introduce a new flag in 1965 a red maple-leaf, already recognised as the Canadian national emblem, was chosen to be the central motif. Australia and New Zealand still have the Union Jack in the upper quarter, or canton, of their national flags. Both also show the Southern Cross, the brilliant constellation visible south of the Equator.

Brazil is another country with tiny stars on its flag. They are said to be in the same positions there as they were in the sky on the night in November 1889 when Brazil declared itself an independent republic.

Some of these and other national flags are shown on the stamps pictured here:

Philippines Republic 2-centavos (1968) *Mexico Olympic Games, 1968.*
Brazil 10-centavos (1968) *Day of the Flag.*
United Nations 3-cents (1951) *U.N.O. flag.*
Switzerland (1944) *Army stamp, soldier of 3rd Division and Swiss flag.*
Antigua 15-cents (1967).
New Zealand 8-cents (1967).
Canada 5-cents (1967) *Centenary of Canadian self-government.*
Yemen 4-bogshas (1968) *Olympic Games in Melbourne, Australia, 1956.*
Yemen 4-bogshas (1968) *Olympic Games in London, 1948.*
U.S.A. 4-cents (1957) *Stars and Stripes.*
Jamaica 3d. (1964) *Flag over map of island.*
Great Britain 1s. 6d. (1969) *Flags of twelve N.A.T.O. countries.*

North Atlantic Treaty Organisation 1949-1969

1/6

Above: At Kolossi Castle Richard I married Berengaria of Navarre, and the Knights Templar and the Knights of St. John of Jerusalem lived here after the Arabs overran the Holy Land in 1291.

Right: Archaeologists are still finding many relics of former civilizations among ruins which date from the Neolithic period. Remains of medieval castles and monasteries are dotted throughout the country.

Below: Cyprus, which has an area of 3,572 square miles (half the size of Wales), is an agricultural country. Wheat and barley are the chief crops and carob beans, fruit and wine are exported. Street traders carry oranges, lemons and tomatoes, and even home-made cheese around on bicycles.

56

Cyprus is a mountainous country, and pack animals are used for transport in the hill districts. For centuries livestock has been a great source of wealth. Cattle, horses, sheep, mules and camels are reared.

ISLAND IN THE SUN

OVER 1,000 years before Christ, Cyprus — the third largest island in the Mediterranean — was a centre of ancient civilisation. The Egyptians were the first recorded conquerors and they were succeeded by the Greeks, Assyrians, Persians and Romans. Richard I of England seized Cyprus in 1191 and sold it to the French. From them it passed to the Venetians and later the Turks. It was bought by Britain in 1878 and made a Crown Colony in 1925. When Britain took over, Cyprus was in decay, but the island was soon on its feet and mining, which had stopped, started again. Today, copper, iron and asbestos are exported. In 1960 Cyprus gained its independence and for the first time in history became a self-governing republic.

Much of the land is still worked by bullock and wooden plough. The soil is fertile but water is scarce. In 1945, a £10,000,000 development scheme was launched to improve farming.

Mexican pilgrims gather every year on 12th December at the Shrine of Guadalupe in Mexico City. They come to pay homage to the Virgin of Guadalupe, patroness of the republic. Right: An Indian in Aztec costume dances for an audience in the street.

Little Miss Mexico shows off her colourful fiesta costume. Dancing will go on all day to the rhythm of drums, rattles and mandolins.

DOWN MEXICO WAY

THOUSANDS of years ago, brown-skinned Indians from Asia settled in Mexico. They built a great civilisation centuries before Christ was born and a culture which proudly claims to have given the New World its first university, its first school for girls and its first printing press.

They built pyramids as grand as those of Egypt. Their wise men knew about mathematics and astronomy and developed a remarkably accurate calendar.

Today, the ruins of ancient pyramids may be seen, and temples which tell the story of a proud race which fell before the onslaughts of European invaders. In 1519, Spanish soldiers, led by Hernan Cortez, climbed the high plateau of Central Mexico, where Mexico City stands today, and conquered the Aztec empire there. For the next 300 years, Mexico was a colony of Spain.

These Mexican villagers are on their way to market with many things to sell. Nothing is sold until the market is reached, however persuasive a passing traveller may be.

Cortez called Mexico City the most beautiful city in the world, and today many people the world over share those sentiments.

Situated 7,415 ft. above sea level, it looks out on a circle of lofty mountains towering above the valley below.

At this high altitude, you would quickly be out of breath if you ran. And your mother would have to boil your breakfast egg for five minutes instead of three because, at this altitude, water has a lower boiling point.

Mexico City lies between two great ranges of rugged mountains, the Western and Eastern Sierra Madres, which rise from the coasts and get higher and draw closer together in the southern part of the country.

Volcanoes

Where the two Sierras meet and become a single range called the Southern Sierra Madre, there is a wild tumble of towering mountains, with the most magnificent snow-capped peaks in Mexico. There is also a trio of very old volcanoes, two of which overlook Mexico City.

It is a city of contrasts, where the ancient, colonial and modern stand side by side. The Aztecs, an Indian tribe, were the first to build there, and a legend explains their choice of site.

The Aztecs were told that when they saw an eagle eating a snake, there they should stop and found their dynasty. One day, on their journey across the country, they came to a large lake with a beautiful island. The priest who led them saw a huge eagle with a struggling snake in its talons. The bird came to rest on a cactus plant and killed the reptile. The Aztecs were overjoyed and set about establishing their city.

There were originally five lakes in Mexico City, and these the Aztecs used to build a city like Venice, with wide handsome streets, half pavement and half canal, so that people could move about either by land or water.

They led a real community life. The older men taught the young boys, who later became priests or soldiers, or were apprenticed to merchants or artisans. There were schools for girls as well. The girls were married between the ages of eleven and eighteen, all arrangements being made by their parents and the priest.

The city was divided into four quarters; in each were people of the same clan with their own representatives who met as a Great Council every eighty days.

When Hernan Cortez arrived with eleven ships, six hundred men, ten cannon and sixteen horses he was so determined to conquer Mexico and send its gold and riches back to Spain, that he burned his ships to prevent his men from returning home.

Montezuma, ruler of the Aztec Empire, allowed the Spaniards peaceful entry into the city, fearing them as white gods. Cortez eventually defeated them and claimed Mexico for Spain.

Spanish dominance remained for centuries in spite of a number of attempts to overthrow it. Not until 1824 was the Republic of Mexico established.

Today, the population of Mexico is about 40 million. About a quarter of the people are Indians and more than half of mixed Spanish and Indian blood. A small foreign population is made up of Spanish, English and Americans.

Mexico City is the sixth largest in the world; a beautiful place with many lovely parks and buildings and gay open markets—and the envy of all other cities in that it has more sunshine than any of them. It is also the largest city not built by a lake, river or sea shore!

THAILAND DANCES

BOARD a magic carpet, shut your eyes and when you open them you might find yourself in Thailand, the land of the Thai (or Siamese) people. Of course, your magic carpet would be a jet plane because Bangkok, the capital, is a big air terminal these days. The Thai are gay, graceful people who love festivals, processions and dancing in the open air. For this, they wear brilliant costumes and often tall, richly ornamented head-dresses. Many of the festivals and some of the dances are based upon old legends. "Thoeng Bong" is the sound which the drum makes, so that is the name they give to the festival dance (seen below). The dancers' movements are made to the rhythmic beat of drums, cymbals, gongs and castanets. At festival time, the canals are filled with beautifully decorated floats.

VENICE OF THE NORTH

The City on the Water, the Capital of Good Design—all these are titles earned by Stockholm, Sweden's capital city and largest port.

Fast water buses in Stockholm harbour carry rush-hour passengers through the city's waterways.

A Swedish liner returns to her home port of Stockholm

FOR the visitor, one impression of the city is lasting. That is that when it comes to design and architecture, the Swedes take nothing for granted. They are prepared to give everything a new and different look—right down to the last detail.

When, in 1908, the City Council decided to build a new City Hall, they created a monumental building in the National Romantic style. The result is a dark red edifice which, although only completed 40 years ago, is architecturally steeped in history, from the exterior design to the last detail of the beautiful furnishings in its dozens of rooms.

Another monument to Swedish ingenuity is Skansen Park—one of the finest open-air museums in Europe. Here, in acres of parkland setting, the Swedes have brought a number of centuries old buildings from all over Sweden and placed them in a natural setting. Visitors wandering through Skansen can see a Lapland woman embroidering at the foot of her raised house, a sixteenth century farmhouse, complete in detail right down to the beehives outside the front door, an ancient church built on stilts to keep it from being snowbound, and many other regional and historic buildings.

Scratch any Swede, the saying goes, and you'll find a nature-worshipper. Nowhere is this more noticeable than in Stockholm. When the first sunshine of Spring comes, Stockholmers never miss a chance to sit out in the parks,

Above: Inside Stockholm's huge City Hall, built in the National Romantic style. Below: Concrete and glass skyscraper office blocks are newcomers to the Stockholm skyline.

Above: The City Hall's roof is built entirely of lead.

Right: A metal sculpture emphasizes the Swedes' love of design, outdoors as well as indoors.

Below: Stockholm's Grand Hotel, one of the most exclusive in Europe.

with their heads back and their eyes closed, facing the sun so as to be among the first to get a tan.

It is this love of natural beauty that accounts for the fact that Stockholm, probably more than any other urban centre in the world, has such a wealth of nature so close by it. The countryside, the woods and the islands just beyond the city limits are reminders that Stockholm is the beginning of an archipelago district stretching far out into the Baltic.

No one seems to be able to agree just how many islands there are in the archipelago. Estimates vary from 25,000 to 35,000. Some are large, with many varieties of flowers and animal life, and others are small bare knobs of rock sticking out of the sea. Some are densely populated, while you may be the first to go ashore on others for several years.

Some are so quiet that the only sounds to be heard are the cries of the

seagulls and the murmuring of the sea. Life seethes on others, like Sandhamn during the international regatta in July. Some have sophisticated hotels and restaurants, while on others the idyllic life of the 'pensionat' goes on unchanged for a century or more.

One of the best ways to see some of these islands is to go by boat to Stockholm. The last 20 or so miles steaming through the bewilderment of green islands as you approach the harbour of the capital is one of the sights of Northern Europe.

The city itself is built on a series of islands and peninsulas; streets threaded with waterways that have earned for Stockholm the name of 'Venice of the North'.

Stockholm did not become the official capital of Sweden until 1634, when it had a population of 9,000. Today 800,000 people live within the city limits and another 300,000 live in its twelve metropolitan suburbs.

Above: A Stockholm park playground, full of colour and unusual design. The climbing apparatus (left) and the slide show Swedish designers' determination to accept nothing traditional.

Left: Another view of the same park shows how perpendicular fountains enliven the scene.

MOUNT TOM PRICE IS COMING DOWN

LANG HANCOCK was the man who found it in 1952. The West Australian prospector noticed that the compass needle of his plane was flickering madly, took a look from the cockpit, and then took another look. Beneath him slumbered a mountain studded with iron ore—a priceless giant which was to be awakened and made to work.

Tom Price was the man who assessed its fantastic potential and persuaded his company, the Kaiser Steel Corporation of America, that it should join with Conzinc Riotinto of Australia to mine this massive mountain of iron which takes his name. Since then, Mount Tom Price, the world's biggest single lump of high grade hematite iron ore, has become the centre of a boom which bears comparison with the early Australian Gold Rush.

From a distance, it looks little different from many other mountains in Western Australia's Hamers-

Hamersley iron ore mine, Mount Tom Price. The mine is situated in the north of Western Australia.

ley Range. Camel-backed in shape, purple-blue in the morning mist, it seems at first just another component of the striking scenery. But closer inspection shows the men and machines burrowing busily into the stockpile of phenomenal wealth.

Here is high-powered industry at its peak. Here are men who end the myth that only faith can move mountains, for it is estimated that within twenty-five years Tom Price will vanish as it is mined flat! When that happens, something like 500 million tons of ore will have been extracted—and even that vast amount will not be all. Other mountains are expected to vanish as they, too, are exploited!

But, if one feature of the territory is slowly going to vanish, another is growing. Until a few years ago, this part of Western Australia was little more than a barren wilderness. Now new towns are taking root and springing up around the perimeter of the twentieth century boom.

The town of Tom Price is a good example of the sort of township this expanding province of Australia expects. There is no question of its being a shanty town, the sort that has always accompanied a boom rush in days gone by. It is not just a saloon, a roughly hewn main street and a general meeting place for the hangers-on that newfound prosperity usually attracts.

It is neat, shining and prosperous. There are brick veneer, three-bedroomed, fully air-conditioned furnished houses, each equipped with such amenities as a washing machine and free water and power. All were expensively built by the major mining company and are

cheaply rented by families who come to the area and are prepared to settle there. There are also an excellent shopping centre, a fine hospital and school, and the widest leisure facilities.

People who are prepared to stay are wanted in this bursting part of Australia. That is why the companies lay on the best of facilities. But it still remains difficult to attract anything more than casual workers.

And, in many ways, it is not difficult to see why. This part of Australia can be like a crucible.

Temperatures on the ground can be 112°, the heat in the digging pits can rise to 150°, and sometimes the machines get too hot to touch. Red dust mushrooms high and coats everything, the noise can be soul-destroying, and the activity is almost incessant.

Perhaps it will settle down soon. The area is still experiencing its growing pains, and it is difficult to say how long they will take to abate. Since 1960, the excitement has been intense and the pace of development frantic.

Until 1960, there had been an embargo in Australia on the export of iron ore, but then it was lifted—conditional on the exploitation of new ore reserves—and geologists began to measure the long-known but undeveloped deposits in the area. Mount Tom Price was among the sources to be tapped, and, by the middle of the sixties, its great iron-ore explosion was in full blast.

RAILWAY MIRACLE

It wasn't easy at first. The recurring trouble with such finds is that they inevitably seem to be situated far from the available means of communication. And it has been the same in this new expanse. Roads and railways have had to be laid, harbours have had to be dredged and deepened to take heavy-draught ships, wharves and loading facilities have been constructed, and water and power laid on.

It is 179 miles from Tom Price to Dampier, the new specially-constructed deepwater port, and laying the railway between the two was just one of the miracles that has taken place. The railway took only 12 months to construct, and that included spanning the large River Fortescue, crossing two sections of the Hamersley Ranges, and building 15 steel bridges, 650 culverts and a 2 mile causeway.

As the expansion continues in this part of Western Australia, the pace of production is increasing. In 1969, 17 million tons of high grade ore were exported from Dampier; which was three times the amount exported in 1967. And as contracts have already been signed to supply nearly 150 million tons to the world's markets, it is obvious that there will be no letting up in Australia's frantic new rush to prosperity.

Western Australia may be hot and uncomfortable at times, but it certainly is not going to be neglected any longer. Some of its mountains might disappear in the coming years, but something else is going to take their place—something like a very comfortable future for an awful lot of Australians.

Rough and ready towns no longer suffice during a boom. The family is considered—and even schools, like this one at Tom Price, are built.

Dampier is Western Australia's main outlet port. From ore loading piers like this, millions of tons of ore can be exported.

COMMATOLOGY IS CATCHING!

By R. K. FORSTER

HAVE you ever heard of commatology? Neither had I until someone told me that this is what my hobby is called. It's another name for postmark collecting.

In 40 years, beginning as a schoolboy in 1931, I have managed to gather 150,000 postmarks from cities and seaports and tucked-away places in every corner of the world.

I began by collecting postmarks from places with unusual names and one of the earliest oddities to come my way was a postmark from Cucumber, West Virginia.

I wondered how a place could come to have such an odd-sounding name and I wrote to the postmaster to find out. He told me that the little town, in McDowell County, took its name from the cucumber trees that grow nearby. These trees belong to the magnolia family and their fruit is shaped like a small cucumber.

From that promising beginning the search went on.

I found there are places in the world with such weird and wonderful names as Sleepy Eye, Rough and Ready, Lucky Strike and Joe Batt's Arm.

Boiling Point, Oregon, was given this curious name because its post office was sited near Emigrant Hill where the old Tin-Lizzie cars used to boil over on a hot day. Total Wreck was the name of a post office established in Arizona in 1881, and Misery is a Swiss village on the Berne-Lausanne rail route.

Postmarks from places like these add a lively touch to one's collection.

In Great Britain alone there are hundreds

of post offices with strange-sounding names.

Bugle is in Cornwall; Trumpet is in Herefordshire. Ham is in Surrey; Sandwich is in Kent. Fortyfoot is the name of a Bridlington (Yorkshire) sub-post office, and near Wakefield in the same county there is a post office called The Fall. Hook and Crook are the names of places in Hampshire and County Durham, and Mumps is a post office in Oldham, Lancashire.

When I went to Mumps to inquire how it got its name I was told it had nothing to do with swollen glands. The name is believed to derive from the old-time "mumpers," or workhouse inmates, who lived in an institution off nearby Cross Street.

Local firms sometimes receive letters addressed in error to Lumps, Humps and even Measles.

HAM, Surrey

Banana, Queensland, is another quaint postal place-name. In the early days transport in this old Australian town was mainly by bullock cart. One such cart was drawn by an old working bullock called Banana "because he was so soft." From this curious scrap of local history the little town took its name.

Jamaica can produce some fascinating postmarks. Among those in my collection are Airy Castle, Bessie Baker, Maggotty, Pepper and Quickstep.

Time and Patience is another Jamaican postmark. To match it there is a place in New Mexico called Truth Or Consequences. This American town was formerly called Hot Springs. In 1950 its citizens decided by a 4 to 1 vote to change the name to mark the tenth anniversary of a popular radio parlour game which had helped to raise funds during World War II.

Locally the town is known as T. and C. but its postmark gives the name in full.

Strange results can be obtained by matching postmarks from places whose names go well together.

Barking, Tooting, Yelling, Clatter, Hammer, Boom, Bangs and Knock are the noisy names of on-the-map places in various parts of the world. Other amusing date-stamp doubles are Black and Blue, Night and Day, Hurry and Scurry and Pains and Relief.

A fine collection of postmarks can be built up quite quickly by obtaining specimens from letters coming to your own home and by exchanging "doubles" with your friends. Once people know you are interested in the hobby it's

SANDWICH, Kent

FLUSHING, Cornwall

surprising how often they can help by sending postcards and letters from out-of-the-way places.

Pen friends are often eager to help and if you have a relative or friend who works in a busy office you should be well on your way to forming a collection if you can persuade him to look out for unusual postmarks on the incoming mail.

Several societies exist for the exchange of material and information about postmarks and many philatelic magazines publish articles about the hobby.

Some collectors file their postmarks in loose-leaf albums. I prefer to mount mine on 6 in. by 3½ in. filing cards. I file these alphabetically, in country order, with a note about the postmark's place of origin on the back of the card.

Among the oddities in my collection are postmarks from Jeddore Oyster Ponds (Canada), the Sea Floor Post Office (Bahamas), Barnegat Lighthouse (New Jersey), a Dynamite Factory

MOUSEHOLE, Cornwall

(South Africa) and the Hotel New Otani (Japan). Many South African National Parks issue special pictorial handstamps and there is even a post office at Multan Central Jail, India.

The stories behind most curious place-names make fascinating reading. Guinea Fowl, Rhodesia, derived its name from the birds of that breed which used to abound in the area. Tomato (Arkansas) was named by a young girl who happened to walk into the town's main store for a can of tomatoes as the name for a new post office was being discussed.

Ravenshoe (Australia) took its name from the tattered remains of a Henry Kingsley novel found in the fork of a tree. Bowlegs (Oklahoma) derives from Billy Bowlegs, an Indian chieftain of the Seminole tribe. Goodnight (New South Wales) can be traced to the nightly greeting of a lonely old shepherd to paddle steamer crews as they plied up and down the Murray River.

Tracking down postmarks from interesting places is rather like going on a personal voyage of adventure and discovery. One never knows where the search will lead or where the journey will end.

And there are bound to be surprises on the way.

GIBRALTAR, Buckinghamshire

MARCH, Cambridgeshire

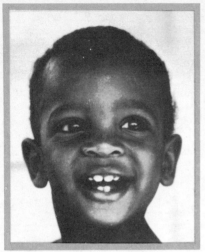

WONDERLAND OF BEAUTY

"**P**LEASURE island of the Caribbean, a land of beautiful beaches, breathtaking scenery and majestic mountains . . . a wonderland of beauty."

This is how the tourist brochures describe the sunshine island of Jamaica.

Although this is not a large country, being little more than half the size of Wales, it is the largest island in the group that makes up the British West Indies. These are part of a long line of islands which curves round, like the inside of a crescent, from the north coast of South America to the south-east tip of Florida in the United States. The sea roughly enclosed by this line is the Caribbean, which was once the home of buccaneers and pirates.

As well as constantly sunny weather, the island has many natural beauties. Mountains cross the land from east to west, the famous Blue Mountains being at the east end. At the foot of this range stands Kingston, the capital of the island, with a magnificent harbour.

Other large towns are Mandeville, Port Antonio, Spanish Town and Montego Bay, a resort of millionaires.

Fine weather and natural beauty bring holiday-makers from the U.S.A. to Jamaica, and wealthy Britons as well. Spear fishing, skin-diving and water ski-ing are the sports which attract them, as well as golfing and horse racing. Young tourists can dance by day on white sandy beaches to the music of a calypso band and, in the evening, at an open-air restaurant or night club. For those in search of entertainment of a different kind, Jamaicans give exhibitions of limbo dancing and fire eating.

Naturally, the tens of thousands of tourists who visit the island each year bring much needed money into the country. By far the greatest number of these arrive from the United States, which is reasonably near, and their shorter journey is not so expensive as the flight from Britain. Nor is Jamaica cheap when holidaymakers get there, for the cost of living is higher than in Britain.

However, the country does not rely solely upon the tourist industry to bring in money. Sugar, rum and bananas are exported. Among other agricultural products, it sends abroad citrus fruits, spices, coffee (Blue Mountain coffee is said to be among the best in the world) and cocoa. Jamaica is also one of the largest suppliers of bauxite, the ore of aluminium.

Equally as famous as the climate is the sunny outlook of the Jamaican people. The population is pre- dominantly African, and to find out why this is so it is interesting to explore the island's past.

Christopher Columbus was the first white man to discover Jamaica which, in the language of its people of those days, meant "The Island of Spring". This was because it was watered by more than one hundred rivers and streams and many tributaries.

When Columbus made his discovery in 1494, the inhabitants were Arawaks of Indian stock. A few years later, the Spaniards colonised the island and their first act was to enslave the Arawaks and force them to mine for gold.

When none was found, the white

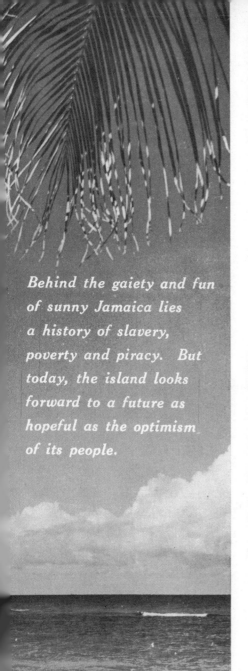

Behind the gaiety and fun of sunny Jamaica lies a history of slavery, poverty and piracy. But today, the island looks forward to a future as hopeful as the optimism of its people.

Sailing is enjoyed by visitors and local people.

The rapids of Martha Brae River provide bathing amid lush, unspoiled beauty.

Jamaica's Blue Mountains are a picturesque background in this suburb of Kingston (below).

men made their new slaves work in the fields. In the days before the Spaniards arrived, the Arawaks had grown cassava (a plant from which bread was made), maize and sweet potatoes. Together with the fish and birds which the Indians caught, these provided their food. In fact, the name, Arawak, means "eaters of meal".

With the arrival of the Spaniards, cattle, hogs and horses were introduced to the island with sugar cane and several varieties of citrus fruits. Cotton and tobacco were also grown; and the wild pineapple, which the Arawaks looked upon as the finest of all fruit, was cultivated.

Heads, not hands, carry the wares in a Jamaican market.

Unable to face a life of slavery, the peaceable freedom-loving Arawaks committed suicide by the thousand; others died from diseases brought to the island by white men.

As substitutes to work the crops, Negro slaves were bought in Africa and shipped to Jamaica, which remained in the possession of the Spanish for 150 years. But in 1654, Oliver Cromwell sent an expedition to the Caribbean to seize a Spanish occupied island and hold it as a base for war operations against other Spanish possessions.

After an unsuccessful attempt to capture Haiti, a neighbouring island, the British took Jamaica, which remained a British possession until it was granted independence in May, 1962.

It did not take Jamaica long to amass considerable wealth. Under a famous Welshman, Henry Morgan, English pirates established themselves at Port Royal on the south coast, at that time the principal town. When they were given letters of marque (official licence to make war on enemy shipping and settlements) they made a number of spectacular raids which brought immense booty back to the island. This they sold to merchants.

Before long, the town was ringed with warehouses bulging with valuable goods which were later sold in Europe at very high prices. As for the jolly pirates, their pockets were filled with coinage from many countries: ducats, doubloons, piastres and the famous pieces of eight.

Another source of wealth was sugar. The English introduced a better cane than the one grown by the Spaniards. Before long, the island was growing so much sugar that there was enough to export to England where, until then, it had been selling at the very high price of 7s. a pound. This was the price of about 80 loaves of bread. Within a century, Britain was importing so much sugar from Jamaica that the wealth of the planters was phenomenal.

The Negro population steadily increased. In addition to the many children born there, more and more slaves were brought in to deal with the sugar harvest. From time to time, some of these escaped and fled to the mountains. There they killed the last of the Arawak men who had settled there, and married the women. These runaway slaves became known as Maroons.

All attempts to capture them failed and in the end, the English signed a peace treaty with them. The Maroons were given 1,500 acres of land and declared to be free people.

The 18th century saw the demand for sugar continually increasing. The second half of the century was particularly prosperous for the white planters; but in England, Christian consciences were turning against the existence of slavery. Led by William Wilberforce, John Wesley and others, an agitation for its abolition was inaugurated. Eventually, in 1833, the local Jamaican government passed a law freeing the slaves.

The joy of the Negro population, then about a quarter of a million, was great; but they were to find their new-found freedom a mixed blessing.

Other countries were still using slave labour, and so they were able to sell sugar and rum at much lower prices than Jamaica could. This was because the plantation owners now had to pay for labour. Also, England refused to allow sugar from the colonies to enter the country at a lesser rate of duty than that paid by foreign countries.

Jamaican exports fell as a result. Planters were ruined and plantations closed. There being no work for the Negro population, the people began to starve.

It was at about this time that the master of a small New England trading vessel visited the island and bought some bunches (or hands) of bananas. These sold well in New

York, and on his next trip he took some to Boston, where they were quickly snapped up. So was born a new trade which grew during the following years until bananas became the island's main crop. Many planters recovered their wealth and even Negro families with smallholdings of two or three acres were able to make a little money from selling their bananas for export.

Until just before the outbreak of the Second World War, the banana trade continued to expand. Then disaster struck. A banana disease known as the Panama disease swept across the island and ruined vast areas of growing fruit. The trade suffered a blow from which it has not properly recovered.

Fortunately, a famous firm of sugar refiners took over some of the ruined plantations and, using new and better methods, converted

In the West Indies, a donkey is a highly prized pet and does many jobs for its proud owner.

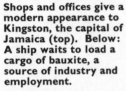

Shops and offices give a modern appearance to Kingston, the capital of Jamaica (top). Below: A ship waits to load a cargo of bauxite, a source of industry and employment.

them back to the growing of sugar cane.

While these changes were taking place, Jamaicans continued to dream of the day when the island would achieve complete independence. They saw that India, Malaya and some African countries were becoming self-ruling, and were not slow to press for the same liberty for themselves.

Other West Indian islands followed their example and formed the West Indies Federation. This became independent on 31st May, 1962; but before then Jamaica had decided to leave the Federation and look after its own affairs.

Its fortunes since independence have been variable.

One can only hope that Jamaica's future will be as sunny as the warmth of its climate and the gaiety of its people.

Many people at St. Mary, Jamaica, work at this sugar factory, for sugar is an important export and earns much needed foreign currency.

A gaìly decked windmill passes on the good news of a wedding to the villagers.

HOLLAND'S 'TALKING' WINDMILLS

HIDDEN radio transmitters were the secret weapons of resistance workers in Europe during the Second World War. But the Germans were aware of these devices and their vengeance on the users was swift and merciless. But despite the vigilance of the Nazis, the resistance movement in Holland was so strong and co-ordinated that messages were successfully passed from one unit to another.

How? The Germans were puzzled! If their intelligence services

These windmills are being used to drive pumps which draw off water from land, reclaimed from the sea, lying 8-12 ft. below sea level.

The sails' positions mean (left) resting, (middle) work beginning, (right), a death.

had looked beyond the level of codes and ciphers and paused to admire the scenery, they might have guessed. Windmills . . . they were the most effective telephones the Dutch had. If only the Germans had realised . . .

Even today windmills are used in Holland to pass on good news about weddings or sad news about deaths. And if you learn their language, you will be able to read their messages too!

A windmill only speaks when it is still, when the position of the sails semaphores the message. Although the windmill's language varies from one part of Holland to another, standard positions are used which have the same meaning in every part of the country.

When the sails are set horizontally and vertically at 90 degrees to each other, it means that the mill, after taking a short rest, will soon start work again.

A diagonal setting of the sails indicates that the mill will be resting for a much longer time.

When the miller wishes to tell all his friends about a birth, a marriage or some other celebration, he sets the sails with the top wing just before the vertical.

Of course, other signs are also used, but these give you the general idea. Naturally, to be of use to resistance workers during the war, a special code had to be worked out, and luckily the Germans never learned what it was.

Towards the end of the 13th century, windmills were used for grinding corn. However, from the 15th century onwards, they were valuable for land drainage and for making polders (reclaimed land units). Grain grown on the reclaimed land was also ground in windmills.

In fact, without windmills, Holland, as we know it today, might be completely different.

Once, 9,000 windmills dotted the Dutch landscape. Today, only about a thousand of these have been preserved, many being historically interesting relics from the past.

Dutch resistance workers in the Second World War used the windmill language to send secret messages to their units.

Wealth hidden in a frozen wasteland was the bargain which America gained when it paid Russia about £1½ million for Alaska, in the extreme north-west of North America, in 1867. America's reward was a rich harvest in duties paid by seal-hunters and the discovery of a wonderland of minerals beneath the ice. Coal, copper, gold, quartz, graphite, platinum, tungsten, chromium, antimony and tin were revealed to the intrepid prospectors.

Recently, this area was being searched for oil—the black gold of modern man—and among the companies which went there was a British one, the B.P. Oil Corporation.

In November, 1968, when the Arctic winter had set in and the ground had become hard enough for large-scale operations, an intensive drilling programme was begun in the Prudhoe Bay area of the North Slope of Alaska, 600 miles inside the Arctic Circle. Four months later, B.P.'s first well in this area, named Put River No. 1, after the River Putuligayuk, struck oil in porous sandstone at a depth of 8,000 ft.

Further tests were required before the potential of the well could be assessed, and more wells were planned to enable B.P. to discover the size of the oil field.

Oil exploration and drilling men are well known for their toughness and their ability to work under almost

ALASKA

A scientist on an observation platform is dwarfed by the massive ice fragments as the *Manhattan* carves a path through the frozen seas of the North West Passage.

any conditions in any part of the world (including on rigs at sea).

But the weather they found in Alaska was probably the worst met by oil-men anywhere in the world. During the Arctic winter, temperatures on the North Slope vary between minus 20 to 30 degrees Fahrenheit, and sometimes it can even be as cold as minus 87 degrees.

In these conditions, great care has to be taken. Oil rigs have to be enclosed against the wind and the cold, and steam hoses have to be used in order to defreeze equipment. Steel becomes very brittle in these conditions, so the heavy drilling equipment has to be handled with great care to avoid breakages.

Because of the lack of roads in the frozen emptiness of northern Alaska, the oil exploration and drilling operations were organised from Anchorage with a forwarding base at Fairbanks Airport for the air-lifting of supplies to the North Slope.

The rig with supplies for Put River No. 1 well was moved into the Prudhoe Bay area in the summer of 1968 through the Bering Straits. But the additional supplies after that, the second and third rigs and the oil men were moved by air, mostly in Lockheed Hercules aircraft. Other aircraft used were Constellations—chiefly as fuel carriers—and Twin Otters for the oil men.

In this ruthless land, flying conditions are very

Ice droplets formed on the faces of the drilling crew.

Land of Hidden Treasures

An Eskimo welder at work on the drilling floor at Put River No. 1 well.

Conqueror of the

Manhattan temporarily stuck in a massive polar floe in McClure Strait. The ice was eventually rammed away.

Put River No. I camp, North Slope, Alaska. This struck oil in porous sandstone at a depth of 8,000 ft. It is 600 miles inside the Arctic Circle.

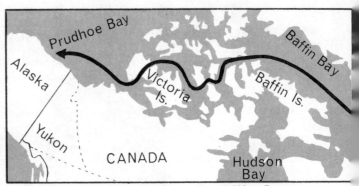

The Manhattan's route through the icy North West Passage.

hazardous. The biggest of these hazards—freezing fog—can reduce visibility to such an extent that the horizon cannot be seen. Because of this, visual direction finding becomes very difficult.

During the long Arctic winter, aircraft can land on lakes which have frozen to the bottom and are strong enough to support runways. In the summer, however, when the lakes have thawed, gravel air strips are laid for the heavier aircraft. The lighter planes, which use skis in the winter, land on floats during the summer.

The severity and complexity of the Arctic weather led to problems in the Prudhoe Bay area which had not been met before by oil prospectors. The chief among these was that of moving the crude oil away from the area. It was hoped that this would be solved by a pipeline laid across Alaska to the nearest warm water port, so that the movement of oil could begin by 1971–72.

Another way of bringing Alaska's oil fields closer, in effect, to the petroleum markets of Europe and the Eastern United States has also been explored. This would mean using the ice-blocked North West Passage which involves crossing from the Atlantic to the Pacific through the frozen Arctic seas.

The conquest by a commercial ship of this passage was made in 1969 when the 1,000-ft. ice breaker-tanker Manhattan smashed her way through the thick Arctic

Frozen Seas

Top: The first load of 48 inch pipe for the trans-Alaska pipeline arrives from Japan. This shipment weighed 6,000 tons. Left: Drillers, air crew, drivers and camp workers take a coffee break at Put River No. I.

ice of the Prince of Wales Strait into Amundsen Gulf and the open waters of the Beaufort Sea. The giant tanker, which displaces 155,000 tons, was the first commercial vessel to negotiate this route. She had a special ice-breaking bow and a helicopter platform on the stern.

On this voyage, the *Manhattan* was accompanied by the Canadian ice-breaker *Sir John A. McDonald* and the American *Westwind*, also an ice-breaker.

Although she was fitted with an ice-breaking bow, the *Manhattan*'s stern, rudder and propeller area often became fixed when heavy ice formed around it. When this happened, the *Sir John A. McDonald* moved in stern first to clear the *Manhattan*'s stern of ice.

All the time, the vessel fought a constant battle with ice, grinding her way through 800 miles of frozen sea. On board were a 54-man crew and 72 scientists, Canadian and American Government representatives and oil company men.

The *Manhattan*'s voyage could lead to the development of huge Arctic deposits of iron, sulphur, copper, nickel, lead and other minerals. World trading patterns would change because of the shorter route, and there would certainly be the biggest ship-building boom since the Second World War.

However, some experts predict that ice-breaking tankers of twice the *Manhattan*'s size will be needed to maintain a regular all-the-year round route.

Geography Picture Quiz

Now that you have read Your Wonderful World, see if you can answer this simple picture-quiz. The answers are on the back page.

1 These people are beside a hot geyser. Where is this and what is the biggest geyser there?

2 Where is this farmer at work? Who were the first recorded conquerors of his country?

3 When did Britain buy this man's country and when did it become a self-governing republic?

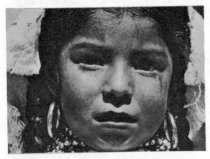

4 This little girl lives in Mexico City. What mountain ranges does her city lie between?

5 What building is this and what is the popular name of the capital of the country it is in?

6 How many islands are said to form the archipelago of which Stockholm is a part?

7 What is the name of the architectural style of the building in which this room is found?

8 A useful mineral is being mined here. Can you give its full name?

9 Who found this mountain in 1952 and who is it named after? Do you know where it is?

10 This woman is taking her wares to market? What is the name of the country she lives in?

11 What were windmills used for in Holland by the resistance movement in the last war?

12 What is this man's job and where does he do it? What is his biggest hazard?

SURVIVAL!

Sentimentality has no place in the laws of nature. One stark truth overrides all others; only the strongest, the bravest and the wiliest survive, as this feature shows.

No animal is without enemies. Some have few, others have many, but all must be ready to defend themselves. They do this by one of two methods. They use either active defence, using whatever weapons they possess, or by passive defence, by some form of armour, or by deterring an enemy without ever striking a blow.

A dog or a wolf uses active defence when it flies at and bites its opponent. Its only weapons are its teeth, especially the long canine teeth, or fangs. But even

81

The tortoise has no teeth and no means of attack, but its skeleton is partly external and it forms a bony box or shell for the animal's safety. The shell is so thick that it is like a mobile fortress into which the animal can quickly withdraw.

A hedgehog (right) has a coat of spines and if it meets an enemy it rolls up into a spiky ball with the head and limbs tucked in. This gives it a perfect passive defence because nothing but an array of sharp spines is presented to the enemy.

these animals will avoid a fight if possible and try, in the first place, to put off an enemy by scaring it. If this is ignored they raise their hackles, bare their teeth and crouch ready for the spring. Their ears are laid back, out of harm's way, the face is twisted into a scowl and the growl turns to a snarl. Only if these signals are ignored do they attack.

Although a skunk's method of defence is very different, there is the same kind of warning. A skunk first stamps its feet, then it raises its tail and waves the tip up and

down. These are signals to an enemy to go away before the skunk squirts a most obnoxious fluid from two glands under its tail. This fluid can be squirted a distance of 12 feet, can burn the skin or hair and cause blindness if it goes into the eyes. Its odour lasts a long time and may spread over a radius of half-a-mile.

The skunk advertises its unpleasant qualities by its prominent white markings, and this has been called a warning coloration. Most poisonous or stinging animals carry

a warning coloration. Usually the colours are black-and-yellow, black-and-red, all black or all red. The wasp is a familiar insect that is coloured black-and-yellow. Its defence is active: it stings. But several insects have a passive defence based on this; they are coloured black-and-yellow, and look like wasps. A young bird catches a wasp in its beak, is stung and thereafter leaves all wasps alone, as well as any insects that resemble them. Such insects are said to mimic wasps: and we speak of this as protective mimicry.

Mobile fortress

Perhaps the best forms of passive defence are seen in the armadillo and the tortoise. In both, the body is enclosed in a bony armour covered with horny plates. The tortoise's shell is more like a fortress into which the animal withdraws. The armadillo has a flexible suit of armour. The hedgehog achieves the same end by carrying a coat of spines and rolling into a spiky ball when attacked.

Some caterpillars have coloured spots on their bodies that look like

The banded armadillo has a flexible suit of armour formed of bony plates, to which horny scales are attached, embedded in the skin. Some armadillos can roll up into a perfectly protected ball as a defence against a predator.

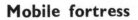

HORNET MOTH   HORNET FLY   WASP BEETLE   WASP

eyes. As long as the caterpillar is crawling normally, a bird may get ready to eat it. But if it draws in its head, thus expanding the "eyes", the bird is scared and flies away.

Many birds depend upon flying to get away from their enemies. However, there are some birds that do not fly, notably the ostrich which has only small wings, incapable of lifting its huge body. When an ostrich fears attack it will lower its head and push its tail up, remaining absolutely motionless so as to merge with the surrounding bushes and trees. If the attacker advances the ostrich relies on its powerful legs to carry it at great speed over the ground. Each leg has two toes, one bigger than the other and bearing a large claw. This is a formidable weapon when an ostrich lashes out at an enemy with its strong legs.

The potoo is a kind of nightjar living in tropical America. It hunts flying insects at night. Its feathers are mottled grey and brown, so that the potoo looks very like the lichen-covered branch on which it rests by day. Its habit is to perch on the end of a broken stump so that it looks like a continuation of it. It keeps its eyes closed so that the bright pupils do not give its presence away.

The bittern, which lives in the reed-beds, particularly in Norfolk, also feeds by night. It is related to the heron but does not have such long legs. It has, however, an equally powerful bill. A bittern will not hesitate to advance on an enemy, with its wings spread and all its feathers puffed out, stabbing with its bill. But it prefers to camouflage itself, and against a background of reeds it is practically invisible.

The way the bittern resembles its background of vegetation is hardly less remarkable than the camouflage used by the dead-leaf butterfly of India. The upper surfaces of the

Some caterpillars, like that of the Fruit-sucking Moth (shown here) have coloured spots on their bodies which look like eyes. When scared by their enemies, the birds, they draw up their heads, making themselves look larger and expanding the "eyes" to frighten the birds away. Many creatures deter an enemy without even striking a blow and use a form of defence which relies upon arousing fear.

wings of this butterfly are brightly coloured, so that when it flies, the gay colours catch the eye. The under-surfaces of the wings are, however, the dull brown of a dead leaf. On the under-surface of each pair of wings is a dark stripe that looks like the mid-rib of the leaf. The veins, usual in insects' wings, complete the resemblance. Finally, the hind wings end in a short "tail", and when the butterfly comes to rest on a twig, this looks like a leaf stalk.

There are two sides to this form of defence. There is the perfect camouflage it provides, and there is also the baffling effect. At one moment there is a brightly coloured insect flying through the air. The next moment, as it settles and folds its wings, the brilliant insect has vanished, leaving only an uninteresting dead leaf.

"Attack is the best form of defence". If animals knew this, it would make a lot of difference. Small birds often fare badly at the paws of a cat, but sometimes a courageous bird, because it has eggs or young, will fly at a cat, calling vigorously, and may end by causing the cat to beat a hasty retreat.

Clearly, the instinct for survival is powerful!

FLATTIES

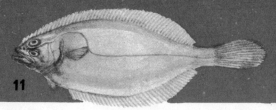

1

2

3

4

5

6

THE life cycle and habits of flatfishes are particularly interesting and only by understanding something of these habits can anglers and professional fishermen hope to catch them.

Generally, people assume that from birth "flatties" such as Plaice and Flounders have broad backs and flat bellies, and live belly-downwards at the bottom of the sea or estuary bed. This is not so. These fishes lay floating eggs which, when hatched out, look like any other "normal" fish.

Only a few days after being hatched, the phenomena of the anatomical change that will make a flat-looking fish out of a normal one begins to take place. The first noticeable difference is that one eye starts to move across the head to the opposite side, giving the young fish two eyes on one side. The mouth of the fish often takes on a different shape and the fins become elongated.

The fish then finds swimming in the normal manner much more difficult and begins to swim nearer the bottom, eventually spending most of its life swimming blind-side downwards across the sea beds.

Not all members of the flatfish family have the same side uppermost. Some are right-sided while others have their eyes on the left side. For instance, the Plaice is a right-sided fish and the Turbot is a left-sided fish.

Many of the 500 or more species of flatfish are quite small, rarely exceeding a few inches in length. A few species are quite large—the Halibut is the largest. growing to almost 8 feet in length.

The Plaice (*Pleuronectes platessa*) is well-known to both sea anglers and housewives. It grows much larger than one would normally expect from the specimens in the shops.

The Lemon Sole (*Microstomus kitt*) is also sometimes called the Yellow Dab. It is considered to be an excellent food fish and grows to about 15 inches long.

The Dab (*Limanda limanda*) is quite a small fish by comparison with some other flatfishes. Mature specimens may grow to 15 inches, but these are not common. The skin of the Dab feels quite rough, which is due to a sharp edge on the scales.

The Flounder (*Platichthys flesus*) is one of the most common sea fishes around the coasts of Europe. Not only are they abundant along sandy shores; they are quite commonly found many miles upstream in rivers.

The large size of the Halibut (*Hippoglossus hippoglossus*) and its delicious taste combine to make this an important food fish but it is never cheap. It is not common in British coastal waters, preferring the deeper waters of the North Sea and North Atlantic.

Another fish with spiny scales is the Long Rough Dab (*Hippoglosiodes platessiodes*). This fish also prefers northern waters where it feeds on small crustaceans, marine worms, etc.

The very rounded head, and the dorsal (top side) fin extending almost to the fish's mouth are distinctive characteristics of the Common Sole (*Solea solea*). Its close relative, the Solenette (*Dugloffidium luteum*) is considered too small to be commercially worthwhile for professional fishermen—it grows to less than 6 inches in length.

Another highly-prized food fish, the Turbot (*Scophthalmus maximus*) has its eyes on the left side of its head. It is scaleless but has blunt, knobbly spines on its uppersides. It prefers the southern coasts of Britain.

The Brill (*Scophthalmus rhombus*) is closely related to the Turbot but is generally smaller. Unlike the Turbot, the Brill has no bony projections on its skin surface. It has a covering of small scales.

Lepidorhombus whiffiagonis is the scientific name for the Megrim. This is a flatfish which can be caught in almost all the deeper waters around Britain's coasts.

The Common Topknot (*Zeugopterus punctatus*) is a rough-scaled, hairy-looking fish whose common name was probably given to it by fishermen.

7

8

1. Flounder.
2. Dab.
3. Common Sole.
4. Long Rough Dab.
5. Halibut.
6. Lemon Sole.
7. Turbot.
8. Solenette.
9. Brill.
10. Common Topknot.
11. Megrim.

9

10

THE PLAICE IS ONE
OF MORE THAN 500
SPECIES OF FLATFISH

Left: the flowers of the potato plant. Right: the seeds.

Potatoes -Our Staple Food

This picture shows the potato plant growing above and below the ground so that you can see the leaves and flowers and the potatoes growing on the roots.

Gold and silver were the attractions which drew Cortez and Pizarro to Mexico and Peru. But the conquerors did not bargain for the big discovery they eventually made . . . the tuber from which our ordinary potatoes are descended. They carried some of these back to Spain and the knowledge of the potato spread through Europe. The English sea captain, John Hawkins, brought the first specimen from South America to England in 1563, and some 20 years later, Sir Walter Raleigh showed some to Queen Elizabeth after the return of the expeditions he had sent to Virginia in the U.S.A.

Cultivation has made the potato one of the most prolific of food plants, for it produces more food an acre than any other crop. It can also be grown in varying conditions, for it matures farther north and at higher altitudes than any other important food crop, except barley.

Several million tons of potatoes are eaten every year in Britain. If you look at the potato plant on this page, you will see that, like other plants, it produces flowers and seeds. But new potato plants are rarely grown from the seeds. Instead, they are usually grown from potatoes which have been allowed to grow shoots and are known as seed potatoes.

The farmer (above) is looking at the seed potatoes from which he will grow his next crop of potatoes. They are ordinary potatoes which are growing green shoots. The shoots sprout from the "eyes" (or brown spots) on the outside of the potato and grow new plants, of which there are about 1,000 main kinds.

Before the seed potatoes can be planted, the soil has to be loosened and prepared with a plough. All varieties are descended from the wild *Solanum tuberosum*, which looks like a cultivated potato, although its tuber is smaller and it produces more seeds.

Planting was once undertaken by people who walked between the furrows with baskets of seed potatoes, dropping them in the earth as they went. Today, men carry out this job with a machine which drops the seed potatoes one at a time into the furrows and then covers them over completely with soil.

Gathering-in or lifting the potato crop can be done by a machine which turns over the soil and brings the potatoes to the surface, where they are picked up. Potatoes are about three-fourths water, the remainder being starch, with small amounts of protein and fat.

A modern farmer gathers in his potato crop with a machine like the one on the left. This lifts the potatoes, shakes off the earth and loads them on to the lorry by conveyor belt. A potato's proteins are in a layer next to the skin, and may all be wasted by deep paring. It is a carbohydrate food.

PLEASE TURN OVER

Potatoes cannot all be sold at once because the demand is spread throughout the year. Those that will not be marketed until later are heaped together and covered with straw and earth to form a clamp.

When they are needed, the potatoes are taken from the clamp and put on to a conveyor belt. This fills the sacks, which are then weighed. Potatoes require a crumbly (not crusty) soil.

The sacks of potatoes are taken in large lorries to vegetable markets in the towns. They are sold to greengrocers and to factories which use them for other products such as potato starch and potato flour.

In this factory, the potatoes are washed and then graded for size. Then they are packed into plastic bags and sent to the shops for sale. The food we eat is the plant's underground store of nourishment.

Potatoes are sold (unpacked) by weight by greengrocers. The starch cells, when cooked, are easily digested.

Large quantities are made into potato crisps, which are very popular for quick snacks; cheese, onion, bacon and even chicken flavours are added.

Millions enjoy fish and chips, well sprinkled with vinegar, bought from fried fish shops, which are very popular throughout Britain.

PLAYING POSSUM

'Dying' is something the opossum does frequently to bewilder his enemies in the hope that they will go away

ANY young American boy or girl who has walked through the woods of the eastern or southern United States and come face to face with an opossum will tell you with delight what happens.

Thinking perhaps that some harm might come to him, "Mr. Possum" does not run away. Nor does he show his teeth and prepare to fight. Instead, he curls up on the ground, half shuts his eyes, lets his tongue loll out, allows his mouth to gape and almost stops breathing.

The idea behind the opossum's strange behaviour in the face of trouble is that he is pretending to be dead.

Of course, any of the opossum's natural enemies, who have seen this trick performed before, are not fooled by it. But opossums seem to think that "passive resistance" saves their lives, because they practise it unfailingly. The trick has given rise to the familiar term "playing possum", which means to ignore or turn one's back on something to which one should be paying attention.

"Mr. Possum," as the American Negroes call this curious creature, is one of the most widely known of American mammals, and is frequently mentioned in the folk-lore and songs of the Negroes of the southern States. This is despite the fact that Negroes are not averse to hunting down opossums with dogs, catching them alive and fattening them up for eating.

The American opossums are the only family of marsupials—that is, pouch-carrying animals—occurring in the New World; all the rest live in New Guinea, Australia and nearby islands. The great marsupial family includes kangaroos, wallabies, koalas and the banded anteater, among many others.

The best known of the American marsupials is the Virginia opossum. He is about the size of a domestic cat, with a body about 20 inches long and a foot long tail which he uses to help him climb from tree to tree. The hair on his body is coarse and he has a pig-like snout and naked ears.

The toes of the Virginia opossum are long, slender and so widely spread that his footprints on the muddy border of a stream or in a dusty trail show every toe distinctly, just like a bird track.

One cannot help wondering how this timid creature, whose enemies in America, apart from Negroes, include great horned owls, foxes, wolves and wild cats, can continue to survive when his only defence mechanism is to feign death—or "play possum".

There are several answers to this question. Firstly, opossums are creatures of the night. They have their dens in hollow trees, in holes under the roots of trees or in similar openings where they can hide away in the daytime. Secondly, during their night-time wandering they will eat almost anything—including a farmer's chickens and any eggs as well.

And thirdly, the reserves of fat that they store under their skins enables them to pass successfully through the long, hard months of winter.

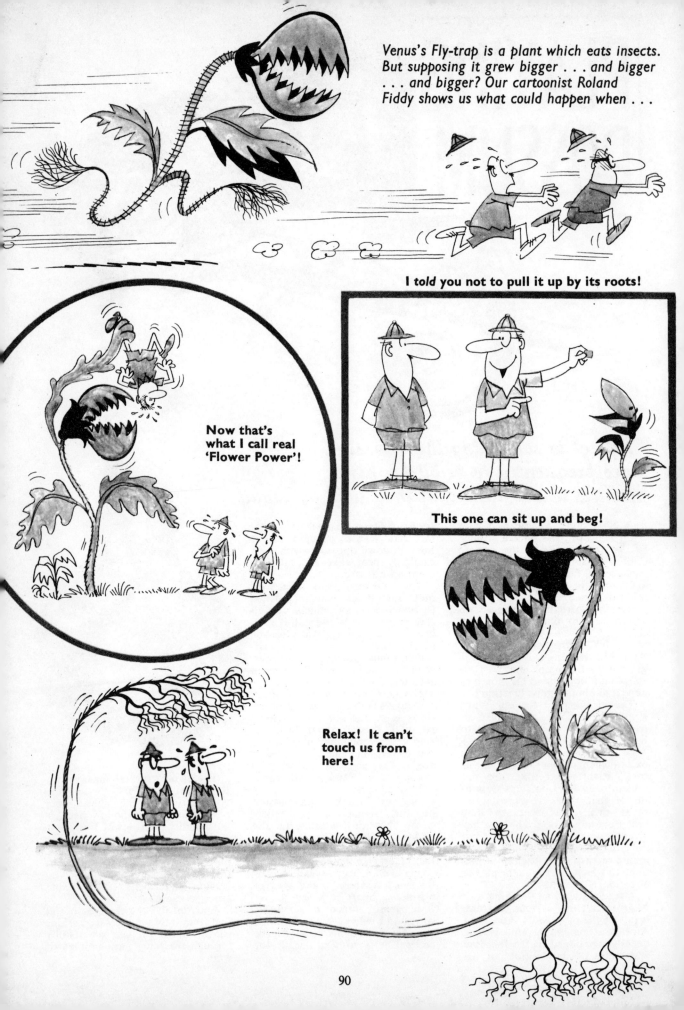

Venus's Fly-trap is a plant which eats insects. But supposing it grew bigger . . . and bigger . . . and bigger? Our cartoonist Roland Fiddy shows us what could happen when . . .

I *told* you not to pull it up by its roots!

This one can sit up and beg!

Now that's what I call real 'Flower Power'!

Relax! It can't touch us from here!

Man meets Plant

Even when mother is running at high speeds, baby kangaroo remains safely tucked inside the pouch.

When baby is only two inches long he needs somewhere safe to sleep. So nature gives him a . . .

MOTHER WITH A BUILT-IN PRAM

A large grey kangaroo gives its off-spring a gentle nudge to persuade it to leave the pouch.

MILLIONS of years ago, some geologists believe, the great continent of Australia became separated from the other land masses of the earth. As a result of this separation, many of the animals found in Australia are not seen anywhere else in the world. One of these unusual animals is the kangaroo.

There are more than 100 species of this likeable and interesting inhabitant of the wide open spaces of Australia, and one can imagine the astonishment of Capt. James Cook and his men, who explored the coastline of the great continent in the 18th century, when they first saw it.

Kangaroos are also found in New Guinea and the nearby islands, but by far the largest number are in Australia itself. They belong to the marsupial order which comes from the Latin word *marsupium*, meaning pouch. Every female kangaroo has a natural cradle or pouch of soft fur formed by a fold of skin below the stomach.

Baby kangaroos are only two or three inches long at birth, although when fully grown a great grey kangaroo reaches a weight of 200 lb. and a length of ten feet from its nose to the tip of its powerful tail. The tiny baby is quite helpless, and would not live for long if it were not protected by the mother's pouch into which it crawls after birth.

The baby remains in the pouch for three or four months until it is strong enough to face the outside world, and then the mother nudges it gently and it pops out to take its first stumbling hops.

A kangaroo's tail is an important part of its body. It uses it for sitting on while at rest, and to give it a powerful extra thrust when it makes its great leaps. A fully grown animal can move across the ground at a speed of forty miles an hour with the help of its tail and immensely strong hind legs.

The front paws are short and comparatively weak, and are used like hands for holding food. When the baby kangaroo is old enough to eat its own food, it is able to lean out of its pouch and nibble grass while the mother is bending over to eat. At the slightest hint of danger, the young kangaroo will leap back into the pouch. If it is some distance away, the mother will rush towards it, gather it up with her forepaws and tuck it into the pouch without checking her flight.

Although kangaroos are timid animals, when cornered they will fight back fiercely, using their front paws to push the attacker down within reach of the terrible claws on their hinds legs which can kill a dog with a single blow.

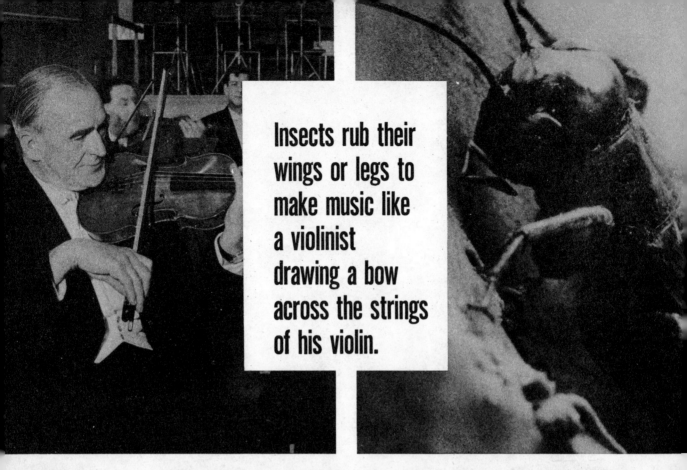

Insects rub their wings or legs to make music like a violinist drawing a bow across the strings of his violin.

THE INSECT ORCHESTRA

THE chirping of a cricket, the whine of a mosquito, the zoom of a passing bee or wasp, even the buzzing of a fly, are not just meaningless sounds. They are the voices of insects talking to others of their own species.

Few people would say that these noises are musical, but they are made in much the same way that a musical instrument produces its notes.

The grass cricket, for instance, might be called the violinist of the orchestra. Along the edges of the cricket's two front wings are rows of microscopic ridges, and when these are rubbed together we get the familiar rasping sound which rises and falls in pitch according to the speed of the rubbing action.

In the same way, long-horned grasshoppers or katydids produce the characteristic call of "Katydid, she did," which gives them their popular name.

Ordinary grasshoppers are also among the raspers, but they use a different kind of instrument. On the back legs there are rows of almost invisible "pegs" and these are used to strike against hard ridges on the wings, setting up a series of vibrations. This produces a sound rather like that made by running a stick along railings.

Rubbing together two parts of the body to make a sound is called stridulation, and this is done by some species of ants and beetles. It is usually confined to the males, in which case it is a courtship call "played" to attract the female. But in some species both the male and female produce the sound.

In certain cases, the sound they produce is of such a high pitch that it cannot be heard by the human ear. At one time, such insects were thought to be "dumb", but by using delicate sound-amplifying devices, scientists can now hear this insect whispering.

An example of a drummer in the insect orchestra is the male cicada. He is often confused with the locust in appearance, but he produces his sound in a totally different way.

In the male cicada's abdomen are two pieces of skin called membranes. Attached to the membranes are muscles which stretch and release them so that they vibrate like the parchment of a drum when struck by a stick.

Close to the membranes are a number of cavities and these increase the loudness of the vibrations.

The Beetles Make Music

Another insect drummer is the death watch beetle. Its tapping noise is made by striking its head against the sides of the tunnel which it bores in wood.

Among the guitarists are the click beetles or elators. The elator has on the underside of the thorax, the section of the body immediately behind the head, a kind of hook mechanism.

When it wants to play, the click beetle bends its prothorax, the rear section of its body, to engage with the hook. Then it pulls with its abdomen muscles until the hook is released with the clicking noise that gives the beetle its name.

If you have a piano, you can pick out the insects' notes with the help of our drawing on the opposite page. This does not show the complete piano keyboard, but only the octaves where the insect notes can be found.

However, the diagram shows the exact pitch produced by some flying members of the insect family.

DRAGONFLY

This insect's 20 wing beats a second produce the musical note E in the fourth octave below middle C.

COCKCHAFER

Strike the note F in the third octave below middle C and you will play the note this insect makes by beating its wings 46 times a second.

LADYBIRD

Now play the Ladybird's note. This is D in the second octave below middle C and is made by 75 wing beats a second.

HORSEFLY

The Horsefly's note is G in the second octave below middle C. It beats its wings 97 times a second to make it.

BUMBLE BEE

You will recognise this one. It is the C below middle C, and is made by 130 wing beats a second.

HOUSEFLY

Now we rise higher to the Housefly's hum; musical note G below middle C caused by 190 wing beats a second.

MIDDLE C

MOSQUITO

A hard worker is the Mosquito who makes musical note D above middle C with no fewer than 307 wing beats a second.

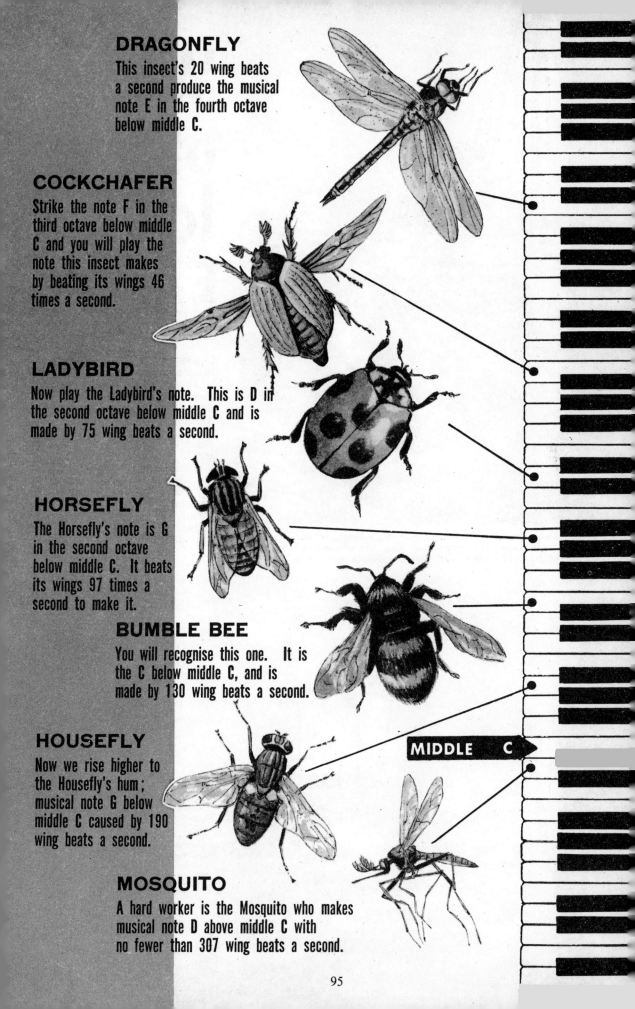

95

Whalers of the Icy Seas

EXCITEMENT, danger and adventure go hand-in-hand in the icy Antarctic seas where whales are hunted for their oil and meat.

Although the harvest is uncertain, despite scientific advances, the catch is always significantly large. In a recent season, 191 catchers and 16 factory ships caught nearly 15,000 blue whales. They could have harpooned more had not unexpected bad weather shortened the season.

From these we get whale oil, which is used chiefly in margarine and cooking fat, but also for soap.

Frozen meat, meat meal (for animals), meat extract, liver oil, liver meal, glandular products (for medicines) and sperm oil are among the things we get from whales.

All of these come from the world's

After being harpooned and killed by the catcher ship (above), whales are towed to a factory ship. If it is not ready to process them, the whales are inflated with compressed air (below) and marked with flags to indicate ownership.

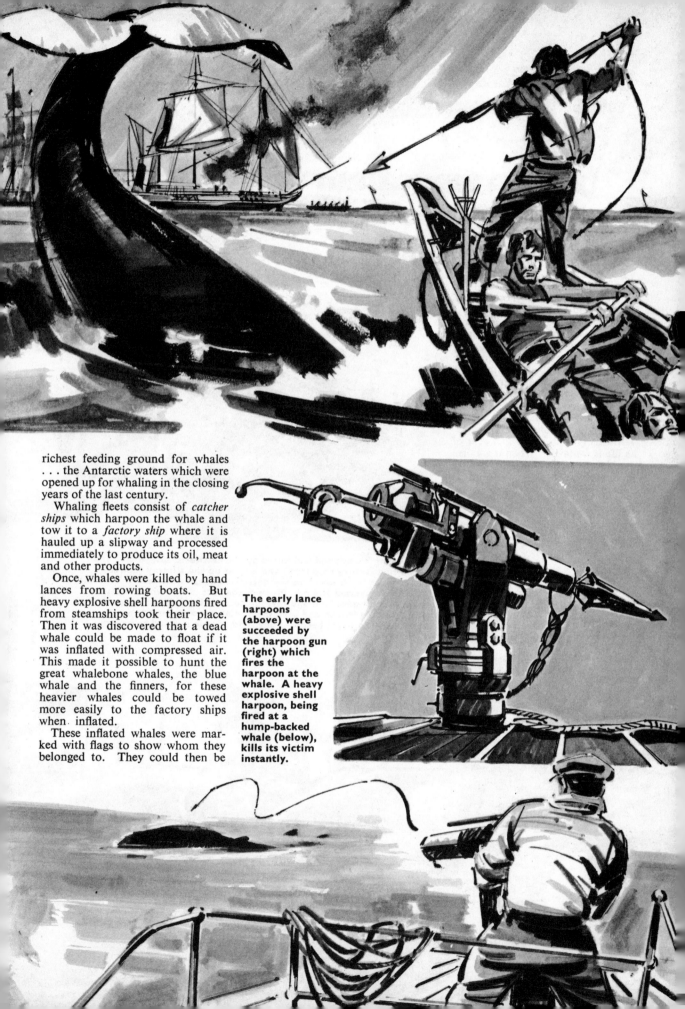

richest feeding ground for whales . . . the Antarctic waters which were opened up for whaling in the closing years of the last century.

Whaling fleets consist of *catcher ships* which harpoon the whale and tow it to a *factory ship* where it is hauled up a slipway and processed immediately to produce its oil, meat and other products.

Once, whales were killed by hand lances from rowing boats. But heavy explosive shell harpoons fired from steamships took their place. Then it was discovered that a dead whale could be made to float if it was inflated with compressed air. This made it possible to hunt the great whalebone whales, the blue whale and the finners, for these heavier whales could be towed more easily to the factory ships when inflated.

These inflated whales were marked with flags to show whom they belonged to. They could then be

The early lance harpoons (above) were succeeded by the harpoon gun (right) which fires the harpoon at the whale. A heavy explosive shell harpoon, being fired at a hump-backed whale (below), kills its victim instantly.

towed to the catcher's factory ship when it was free to process them.

If the catch is a Blue Whale, it is certainly a prize worth having. It is the greatest and most highly valued of all whales.

It is fortunate for us that this whale is a sea animal for its gigantic size alone would make it a destructive force on land. Prehistoric animals possessed enormous bulk, but an adult blue whale is three times larger than the biggest of these. After eleven years of life, it grows to a length of 100 ft. and weighs as much as 100 tons.

But its growth in the first year of its life is fantastic. When it is born, the blue whale is 24 ft. long and in the first seven months of suckling it puts on weight at the rate of 10 lbs.

After being harpooned and killed by the catcher ships, whales are towed to a factory ship (top) and hauled up the slipway (above). The blubber is removed on the deck with flensing knives (below) and is boiled to extract the oil. Much of the rest of the whale is also used. The oil goes to make margarine, cooking fats and soap. Some of the meat is frozen and used for human consumption and the poorer quality meat becomes dog food. Some is dried to become cattle meal.

an hour! When it is weaned, it is 54 ft. long and weighs about 45 tons.

To the whalers, this represents about 23 tons of whale oil.

The fin whale or rorqual is the blue whale's more numerous relative, is equally important to Antarctic whalers, and grows to 90 ft. long.

During our winter, which is the Antarctic summer, the blue whale and the finner go to the Antarctic pastures, and both are thin when they arrive after a long fast with their young in warmer but comparatively foodless seas.

However, they soon put on weight and yield a rich harvest. It is not uncommon for more than 7,000 blue whales, 18,000 fin whales, 2,000 humpbacks and 400 sei whales to be caught in a season.

But the catch in the early days was minute compared to this. A 1906 expedition led by Captain Christiansen had a factory ship which was not specially built for the purpose, and the whales had to be cut up alongside from rafts or platforms.

Six years later, after phenomenal progress, the world's whaling fleet had 41 floating factories, of which 37 were Norwegian.

Whales are hauled aboard a factory ship, for treatment, up a stern ramp or slipway. The first factory ship to be equipped with one of these was the *Lancing*, a Norwegian vessel, in 1925. This was the first ship to process a whale outside territorial waters. Before that, Britain owned all the harbours in which a factory ship could do its work in shelter.

So swift was the progress which followed that, until the Second

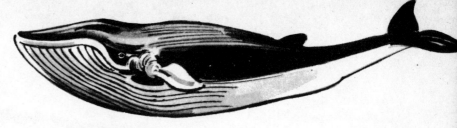

FIN WHALE

HUMPBACK WHALE

SPERM WHALE

SEI WHALE

Whales can be found underwater by an echo device. An oscillator fixed below the ship sends out pulses of sound, too high to be heard by the human ear, in a narrow horizontal beam. These are reflected by objects, such as a whale, in the path of the beam and give accurate information as to their position. This gives the harpoon gunner a "sixth sense". Once the whale has been "sounded", the catcher ship patiently follows it until the gunner can take his shot when the animal surfaces. Afterwards, the whale is towed to the factory ship.

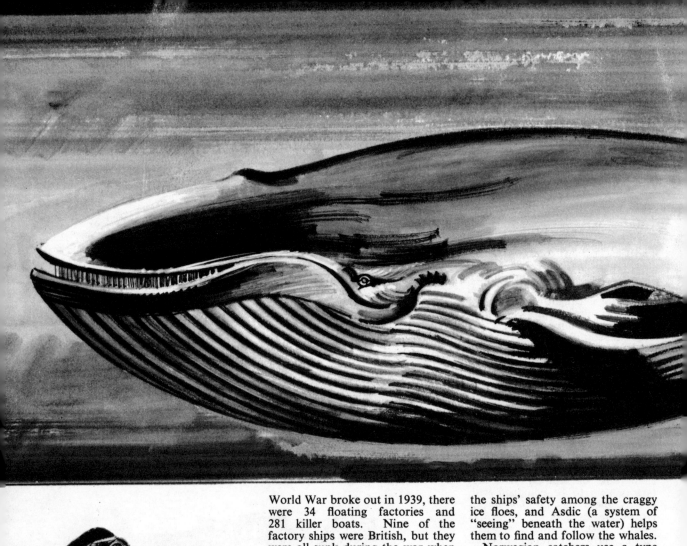

World War broke out in 1939, there were 34 floating factories and 281 killer boats. Nine of the factory ships were British, but they were all sunk during the war when they were used as transport ships and troop carriers.

Heavy losses were also suffered by the other nations. Replacements have been built since. These new factory ships are larger than their pre-war counterparts and operate with 12 catchers.

Electric aids help in locating the whales. Radar is used to ensure the ships' safety among the craggy ice floes, and Asdic (a system of "seeing" beneath the water) helps them to find and follow the whales.

Norwegian catchers use a type of Asdic which vibrates sound through the water which frightens the whale. This makes the whale swim away quickly in a straight line so that he can be overhauled and caught.

These three prehistoric monsters were giants of millions of years ago. Edmontosaurus (left), Brontosaurus (middle) and Tyrannosaurus (right) were the largest of the flesh-eating dinosaurs. But they were midgets beside the Blue Whale.

Once this was done with a hand lance, which was followed by the explosive harpoon. An idea to follow these is the electric harpoon which paralyzes the whale when it strikes him and kills him very quickly.

Is there any danger that such wholesale harvesting will make the blue whale and its fellows extinct?

With limitations on the number to be caught each season, it is to be hoped that these fascinating creatures of the deep will continue to survive for generations to come.

The Blue Whale (above) is as big as three dinosaurs (left), 17 elephants or 133 oxen. It is the largest animal the world has ever known, even in prehistoric times. One weighed 163 tons and another 118 tons.

ELEPHANTS— The Gentle Giants

A three feet high animal with tiny pig-like ears, small tusks and just the beginnings of a trunk was the *moeritherium*—the elephant's earliest ancestor . . . 60 million years ago. Through the ages, the elephant developed through the mastodon and the mammoth to the two kinds we know today, African and Asiatic. These have several differences, but the African elephant is usually recognised by its larger ears. Man has long used these creatures to help him in his work. In fact, as long ago as 2,000 B.C., they were harnessed and made to move things too heavy for man, as the ancient coin on the right clearly shows.

Alexander the Great (356–323 B.C.), King of Macedon, and the most famous figure in all Greek history, was the supreme military genius of the Ancient World, who won battles against many nations. He explored and conquered parts of India where he rode his horse, Bucephalus, against an enemy mounted on elephants. It was here that the horse died of exhaustion and old age in a battle. Alexander founded a town nearby in its honour and called it Bucephala. The Indian elephants he encountered in this campaign had been tamed by being driven into a small enclosure and cut off from their herd by other tame elephants.

Through a mountain gorge marched an amazing procession of men—and strangest of all—elephants! This was the army of Hannibal, the great general of Carthage in North Africa, passing through the Swiss Alps on their way to attack Rome in 218 B.C. Although ice-cold storms of snow and sleet lashed them, Hannibal's army did not falter. Only one elephant (Hannibal's) survived the invaders' victories. Years later, in 202 B.C., Hannibal was beaten by the Romans.

Although Julius Caesar visited Britain in 55 and 54 B.C., it was not until about a century later in A.D. 43, that Rome began the conquest of the island. Roman soldiers crossed from Gaul and within three years had conquered it as far north as the Humber and as far west as the Severn. They are said to have brought elephants (right) to the island, which they ruled for nearly 400 years.

African elephants are hunted for the ivory of their tusks, which is made into beautiful carvings. Ivory once had many uses, but plastics have replaced it to a large extent.

Asiatic elephants have long been captured and domesticated and used as transport animals in India, Burma and Thailand. With its mighty trunk and tusks, the elephant can lift huge logs and carry great weights. Elephants do not breed easily in captivity. But once caught and tamed, they soon learn to obey.

The only elephants most of us see are those in a circus (right) or a zoo. Their gentleness, docility and readiness to obey commands make them firm favourites with children and adults. Crowds in Asia love them, too, where decked out in glorious finery they walk with majesty and dignity in magnificent parades (above). Truly man has found a friend in these trusting, obedient beasts.

103

The Nature Picture Quiz

Now that you have read the Nature section, see if you can answer this simple picture-quiz. The answers are at the back of the book.

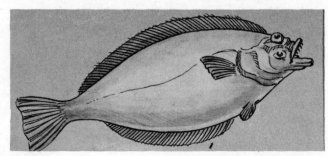

1 This animal has a coat of spines and if it meets an enemy it rolls up into a spiky ball with its head tucked in. What is its name?

2 This species of flatfish grows to a great length. In fact, it is the largest flatfish. Do you know its name and how long it becomes?

3 Before they are needed for sale, potatoes are stored. What is the name of a potato store?

4 This is a neglected part of the potato plant. What is it?

5 What is the name of this animal which pretends to be dead when it is afraid?

6 Where would you have to go to see this animal in the wild? What is it?

7 This insect's sounds are made in the same way as a violin's. What is its name?

8 Do you know how many times a second this insect beats its wings? What is it?

9 Which breed is sought as the greatest and most highly valued of all whales? Where is it caught?

10 You will recognise these elephants, but what are the names of the countries in which they live wild? Who rode his horse against an enemy mounted on elephants?

It was said of the American writer Jack London that his best book was his own life. Here is the story of this . . .

SAILOR WITH A FIERY PEN

Jack London was a prolific writer, with a sparkling imagination and an exceptional capacity for work, so much so that in only sixteen years he wrote more than fifty novels. In these he describes all different types of surroundings and his characters belonged to every strata of society. From the Polynesian islands to the frozen expanses of the Klondike, from the most savage lands to the most civilised cities, a whole glittering world lives and moves in his books . . .

ON an August night in 1913 a searing fire lashed the skies above the Sonoma valley, an area of fertile ranchland above San Francisco.

The centre of the blaze was Wolf House, a dream house built of huge redwoods and massive red stones. It had taken three years to build, and cost thousands of dollars.

The owner had never lived in it, and now he stood watching his dream destroy itself.

His name was Jack London, and all his life he had wanted a place of his own. Now it was gone. He never rebuilt it.

We cannot know what London thought as he watched Wolf House die. But certainly some memories of his tumultuous life must have passed through his mind.

And perhaps his thoughts turned back nearly 20 years to the freshman class of Oakland High School in Southern California on the first day of a new term.

The other children in the class tittered at the new boy who took his seat among them. For one thing, they were all children from good homes and they wore clothes that were clean and neat, while the newcomer was in a crumpled suit, he had no tie on, and he was chewing tobacco.

And for another thing, while the average age of the class was fourteen and a half, the new boy squirming his big frame uncomfortably on the hard seat, was very old indeed to the other pupils—he was 19.

The boy was Jack London. His name meant nothing then to his classmates, nor to anyone except his family and a few friends.

But when he died, 21 years later, his name was celebrated round the world as one of America's great authors.

But his fame was still a few years ahead. At the moment, his ambition was to enter the University of California, and for that he needed more schooling than he had so far received in his crowded life.

Already at 19, he was a mature man in many ways. When he was only 15 he had been an oyster pirate—slipping out in his sloop Razzle Dazzle in the dead of night to steal oysters from the beds in Lower Bay.

Tiring of this, he shipped out before the mast in an 80-ton schooner bound on a sealing expedition to Japan, gaining at first hand that vigorous knowledge of seafaring life which was to bring him so much fame when he came to write it down.

Back home in Oakland he took a job shovelling coal in a power station. It was grim work. Life in America for poor people in those times was a hard business and London was a poor man, he had his mother and sister to support, and he always managed to raise money for them.

Only one thing supported him in those dark days: a growing love of books, fostered by a friendly librarian in Oakland.

One day he noticed that one of the local newspapers was running a short story competition. London sat down and wrote an account of a typhoon raging off the Japanese coast. His story won first prize, and he decided that from now on writing would be his career.

But it was not easy. Always he and his family were on the brink of ruin. He was a familiar figure at the pawnshop.

When the job shovelling coal finally exhausted him he became a hobo—hiding away on freight trains as they thundered across the length and breadth of America. For a year he lived this way, but when he came back again to Oakland he realized that he did not want to be a wandering labourer all his life.

He wanted to be a writer, but he wanted to go to the University first. But for this it was necessary to pass examinations, which was why he found himself among the giggling teenagers in the freshman class.

London stuck it for the required three years, and then entered the University of California. But what can a university, which is rightly regarded as a preparation for life, teach a man who had already done so much? London never finished the course.

His next destination was Alaska. Gold had been found in the

When he was only just sixteen Jack London was the owner and captain of the "Razzle-Dazzle", a little boat which he had bought only by making great sacrifices. With this boat and the help of a young girl he devoted himself to clandestine oyster fishing: a difficult, dangerous and illegal occupation.

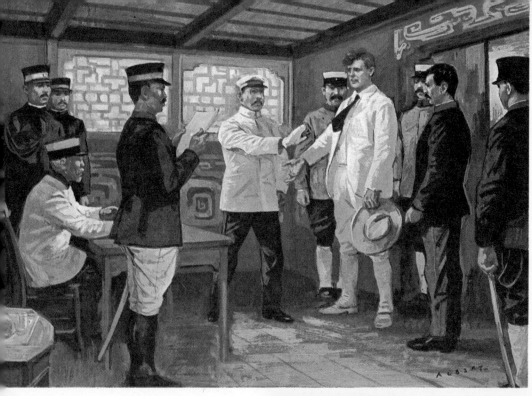

In 1904, when he was named as war-correspondent in the Russian-Japanese War, Jack London had to appear before a military tribunal. London had reprimanded and slapped a servant who had been stealing. This episode provoked harsh accusations against London from which he had great trouble in clearing himself.

On the 18th April, 1906, whilst he was in San Francisco, Jack London witnessed one of the greatest catastrophes which have ever afflicted mankind: the great earthquake of San Francisco.

Klondike, and London wanted to share in the adventure. Of this there was plenty, and though he came back from the Klondike penniless he had ample material for the short stories that he now started to write in earnest.

One day he sat down and wrote a novel about a dog—someone's pet—which was kidnapped and taken to Alaska to help draw a sledge in the Gold Rush. The story was *The Call of the Wild*, and it made London world famous.

This was his moment, and from now on all the adventures he had known poured from his pen. The terror and wilderness of Alaska, the fury of a South Seas storm lived vividly in his readers' minds. He became a hero to the public, because here was a writer who not only wrote about exciting things, but did them for himself.

One of London's most famous realistic books was *The People of the Abyss,* which dealt, not with poverty in America, but here in London. He found himself with time to spare in England, and instead of living in a grand hotel, as he could well have afforded, he lived in a dreadful hovel in the East End of London.

But all his life, one of London's great dreams was to have a real home of his own. His penniless childhood had been spent moving from place to place. Wolf House was to mark the end of his wanderings. But, on that August night, it burnt down. It was a blow which hit him very hard.

He began to drink heavily, and took his own life in a fit of despair in 1916, at the early age of 40.

The Play could last all Day!

This actor in a Nô play is masked, a tradition in this form of play which dates back to the 15th century.

IF you go to the theatre in Japan, it is just as well to take your lunch with you. The performance may last all day.

The audience squat patiently on their heels for hours on end. In the brief intervals, they produce cold food and chopsticks.

There are three types of play in Japan—sacred, classical and puppet. They are quite different from Western plays. Women seldom appear, their parts being taken by men.

In the sacred, or *Nô*, plays, the actors are masked. This is the highest form of play, dating back to the 15th-century.

Originally, the *Nô* plays were given in front of temples. Now they take place on a small square stage with a bridge at the back. The actors wear wonderful carved masks and silk costumes, using fans and big sleeves for effect.

The gold background with a painting of a crooked pine tree is unvarying. There is a chorus and orchestra of flute, tambourines and samisen (a three-stringed guitar).

The strange performance combines sacred dances, poems and singing. It is so slow that it is interrupted now and then by deliberately comic dances called Kyogens. The Japanese believe *Nô* is perfect art.

The easiest type of Japanese play for Westerners to understand is the *Kabuki*.

For this, the stage is enormous with elaborate detailed scenery. Dances, love, war and history are presented by up to fifty actors.

The actors enter the stage from a gangway bridging the theatre to the sound of drumming or flutes. The drumming is made by striking a wooden board with a mallet.

At the end of each act, scene-changers in medieval costume run across the stage with a curtain to fast drumming. They retrace their steps to slow flute music.

Kabuki plays are so constructed, with the gestures so expressive, that each event is a perfect picture.

The third type of play is the puppet-play show called the *Bunraku*. This became popular when actresses were banned from Japanese stages centuries ago.

However, actresses have begun to appear in plays in recent years.

The gold background with a painting of a crooked pine tree is unvarying in *Nô* plays (left). You can see the chorus and the orchestra of flute, tambourines and samisen (three-stringed guitar).

Actors in Japanese plays wear wonderful silk costumes (right) and use fans and big sleeves for effect.

Performances combine sacred dances, poems (below) and singing. Now and then, comic dances are introduced.

Puppet plays in Japan feature figures with skilfully designed head and hands (above), and each puppet has three handlers.

Henry VIII saw Holbein's beautiful portrait of Anne of Cleves and decided to marry her. But Henry had a shock when he found out what Anne was really like.

Peter Jackson

PORTRAIT THAT FOOLED A KING

A PAINTING by Hans Holbein the Younger was the unwitting cause of one of King Henry VIII's disastrous marriages. King Henry had divorced his first wife and beheaded his second wife. His third wife, Queen Jane, who had been queen for just a year and a half, had given Henry what he most wanted, a son to rule after him, but soon afterwards she died.

Henry was overjoyed at having a son. It meant that the Tudors would go on ruling England through his son and the Kingdom would continue to be at peace. With an heir to the throne, there would be no cause for civil war when Henry died, with rivals for the throne fighting each other and bringing misery to the country.

Now King Henry decided to look for a fourth wife. He thought of a French princess and asked King Francis of France to send all the fairest ladies of his court to Calais, so that he could choose one to be his wife, but King Francis replied that he could hardly send the ladies of his court to be inspected like horses at a market.

Henry's chief minister, Thomas Cromwell, thought it would be better to marry a German princess. As King Henry had made himself head of the church, instead of the Pope, Cromwell was afraid that the Catholic rulers of Europe might join together against England, urged on by the Pope. The princes of Germany were Protestants, so they also did not consider the Pope to be the head of their churches. For this reason, Cromwell thought that Henry should marry a German princess and join up with these princes. He chose Anne of Cleves, the second daughter of the German Duke of Cleves.

Anne's eldest sister, who was married to the leader of the Protestants in Germany, was beautiful, charming and clever and Cromwell was sure that Anne would be the same.

Just to make sure, King Henry ordered the famous portrait painter, Hans Holbein, to paint a portrait of Anne for him to see and in case Anne was not to Henry's liking, he was to send a portrait of her younger sister, as well.

After studying the portraits, Henry decided to marry Anne and he was so impatient to see his new bride that when she finally arrived in England, he rode down to Rochester in disguise to see her.

His first meeting with Anne was a terrible shock. It was clear that Hans Holbein had done his best to flatter her in the portrait. Not only was she plain, but smallpox had left her face pitted with ugly scars and Holbein had deliberately left these blemishes off the portrait.

Probably, Anne was just as disappointed with her first sight of Henry, who was middle-aged and growing very fat.

In fury, Henry sent for the men who had arranged the marriage and told them that they must find some way out, for he would not marry Anne.

There was nothing they could do, the wedding was arranged and the unwilling Henry had to marry his ugly bride.

It was too much for Henry. He sent his chief minister, Thomas Cromwell, to the Tower, accused him of treason, and had him beheaded. Then he fell in love with the young and beautiful Catherine Howard and decided to make her his fifth wife.

Anne of Cleves was told that if she would agree to have the marriage dissolved, she would be treated as the

This is the portrait of Anne of Cleves commissioned by Henry VIII. Holbein created a beautiful pattern but not a true likeness.

Holbein painted this portrait of himself. Some art experts feel that his style suggests Italian influence.

Holbein flattered his sitters, as can be seen from the sketch of Anne of Cleves (above). He also symbolically represented a man's calling, as in The Merchant George Gisze on the right. The merchant sits sharp and watchful amid the trappings of his trade.

king's sister and given an estate and a large income. Anne was quite willing and the French ambassador said, after the marriage had been dissolved, that she now looked much more joyful than she had before.

Holbein, whose portrait had caused all this trouble, was born at Augsburg in 1497. His father, who was also a painter, is usually known as Hans Holbein the elder.

Holbein received his first art lessons from his father, but in 1515, he left Augsburg and went with his brother, Ambrosius, to Basle, where it is thought he worked as an apprentice to the artist Hans Herbst.

It is probable that Holbein's first patron was the famous scholar Erasmus. At any rate, shortly after arriving in Basle, he made some pen-and-ink drawings to illustrate a book by Erasmus called *Moriae Encomium* ("The Praise of Folly").

At about the same time, Holbein designed title-page blocks for new editions of the Bible, and for some works by classical authors.

Two of Holbein's earliest portraits are those of Jacob Meyer and his wife, now in the Basle Museum. These are thought to have been finished in 1516.

In 1517, Holbein went to Lucerne and was employed for a while in the house of Jacob Hertenstein. The portrait of young Benedict Hertenstein, now in New York, belongs to this period.

Holbein remained only two years in Lucerne, however.

At the end of that time, he returned to Basle and it is thought that he then married.

Some people believe that before 1519—when he returned to Basle—Holbein may have gone to Italy There is no positive proof of this, although his style suggests considerable Italian influence.

Holbein came to England at least twice during his lifetime.

The first occasion was in 1526. Erasmus provided him with letters of introduction to Sir Thomas More, an eminent statesman of Henry VIII's time.

Sir Thomas More treated him very kindly and even entertained him at his own house in London. While he was there, Holbein probably made the sketches for a group representing Sir Thomas More and his family. This picture is no longer in existence.

There is, however, a drawing of the head of Sir Thomas More by Holbein which is now at Windsor.

In 1528, Holbein returned to Basle. With the money he had been able to make in London he bought a house for his family. Portraits of his family belong to this period.

At this time he also painted a portrait of an unknown young man, wearing a large soft hat and some Bible illustrations known as *Icones*.

Holbein is thought to have contracted the plague in London where he died in 1543.

THE BUILDERS OF BRITAIN

We are lucky to have in Britain a rich heritage of architecture. Nearly every village or town with a history going back a hundred years or so will contain several examples of different styles of architecture. These may be seen in the design of private houses or be typified by a parish church. As you look at these, you will realise that the growth of architecture in Britain, as through the ages, it passed from one style to another, can be viewed as an example of European building as a whole, which is the most diverse in the world.

Although British building closely followed the European pattern, it has always borne a touch of distinction. The flowing tracery of Decorated Gothic windows, called the Flamboyant Style, and the richly decorative fan-vaulting of the late Perpendicular Gothic period, are thought to be of British origin. Also, the stately double-towered entrances to large houses and colleges of the Tudor age are only found in these islands.

The aim of this feature is to help you to recognise some of the characteristic features of the buildings you may visit. We begin with the simple buildings of pre-Roman times.

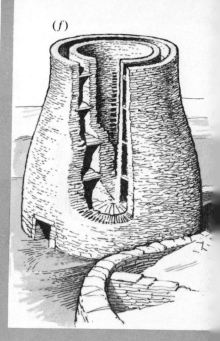

Above: A late Neolithic hut (a) is made of branches tied at the centre, and a dry-stone wall around the base. The Iron Age hut (b) has a turf and thatch roof and wattle and daub walls. Earth banks anchor the tent-shaped hut (c), and (d) is a dry-stone beehive hut. Iron Age people went through a small passage into the stone hut (e) which has a thatch roof. On the right (f) is an Iron Age Broch Tower, a village fortress, about 40 ft. high.

A simple Neolithic dolmen (g) contained the remains and utensils of an important person who had died. The stones were covered with earth and turf (h).

Stonehenge in Wiltshire (i) is the most famous of British megalithic stone circles. It was put up in about 1800 B.C. in late Neolithic times and improved in the early Bronze Age. Its purpose is unknown, although it later came to be used by the Druids or ancient priests.

K

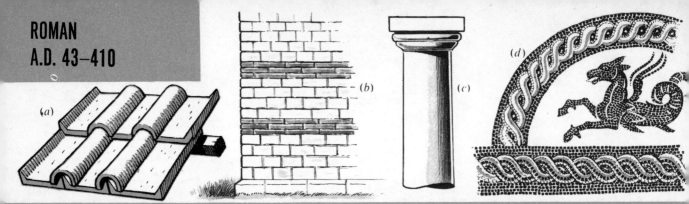

ROMAN
A.D. 43–410

Many architectural survivals of our 400 years of Roman rule exist today. Roman roof tiles (a) were made of clay. The Roman tile bonding (b) goes right through the thickness of the wall. The Roman style of Doric capitol (c) and the mosaic flooring (d), which was used extensively in villas, also appeared in Britain. Below are (left) a Roman mile fort on Hadrian's Wall, a Roman villa (middle) and (right) a theatre, sometimes cut into rock.

ANGLO-SAXON
A.D. 410–1066

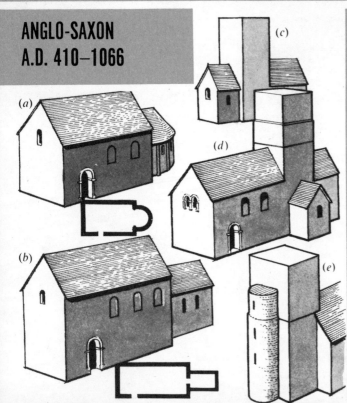

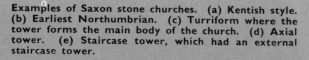

Examples of Saxon stone churches. (a) Kentish style. (b) Earliest Northumbrian. (c) Turriform where the tower forms the main body of the church. (d) Axial tower. (e) Staircase tower, which had an external staircase tower.

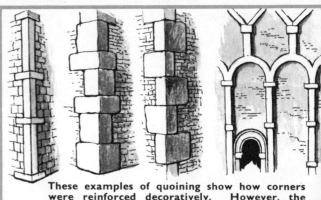

These examples of quoining show how corners were reinforced decoratively. However, the Pilaster work on the right was used for appearance only.

The pictures here (f, g, h and i) all show the different styles of Saxon windows. On the left is a typically rounded arched door (j) with continuous strip work around the outline. Churches of the Saxon period were not made of stone until the 7th century, wood being used before this time. Even after this, ordinary houses were for long made of wood.

NORMAN
A.D. 1066–1160

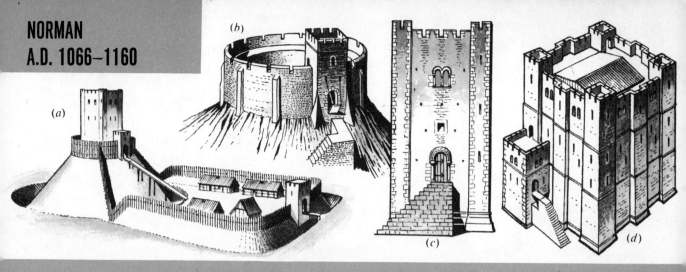

Norman rule brought many castles to Britain. Many wood ones had a motte and bailey layout (a). Shell keeps (b) were small circular enclosures open to the sky with stoutly defended entrances. These had followed simple stone strongpoints called keeps (c). In later Norman times, bigger hall keeps with extremely thick walls were built (d). These had defended entrances into the first floor. In many cases, a dividing, vertical, internal wall was constructed to give more comfort to the inhabitants. Many castles like this still exist.

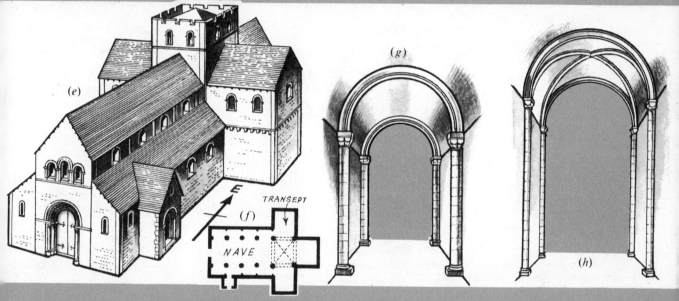

Church architecture : The plan of the church, abbey or cathedral was now in a form of a cross with the transepts making the arms (f). Churches had thick walls, narrow aisles, small windows high up and, generally, squat towers (e). Vault construction was either of the simple barrel type (g) used by both Saxons and Normans, or ribbed vaults (h) when a further two arches were sprung and crossed diagonally between each two crosses. Below: Typical Norman many arched doorway (i). Window with a decorated drip stone course and string course (j). Massive and stumpy nave piers (k). Example of interlacing arcading (l) used for ornamentation. Norman billet design (m) and Norman star decoration (n). Chevron ornamentation (o).

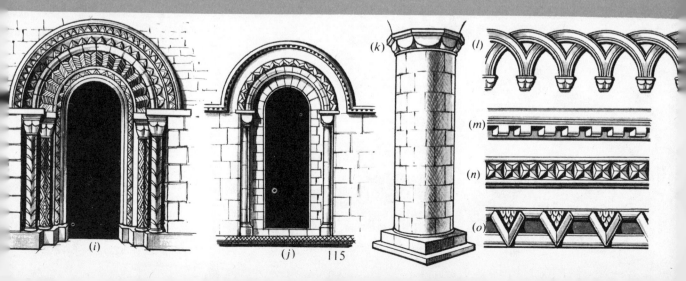

115

EARLY ENGLISH GOTHIC
A.D. 1160–1275

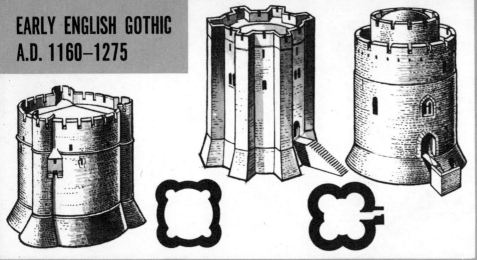

The advent of the Crusades and travel to and from the Holy Land by engineers and architects brought the growing influence of the Saracen round tower in castle building to Britain. The old square towers were easy to batter down if blows were aimed across the angle of the corners. Battering a round tower only pushes the stones closer together. The three examples (above) and the two plans give an idea of the theme and variation of a round keep.

Stoutly defended, double-towered entrances became the fashion, as can be seen above. Castles were expanded into an inner and outer bailey system.

(a)

(b)   (c)

(d)

Early English gothic is the purest of the period. Above is a simple twin window opening with a quatrefoil opening above all, contained in a unifying pointed arch (c. 1230). This forms a feeling of spiritual perfection.

A simple three-lancet window of the period (c. 1200).

Here is a later early English window opening of about 1260. This period witnessed the foundation and growth of a great many monasteries' of different orders. They were, in fact, found in great number between 500 and 1500.

(a) Early English arcading. (b) Trefoil. (c) Crockets. (d) Decorative ornament. You may still see examples of these when you visit an ancient church.

AN EARLY MONASTERY

1. Gatehouse. 2. Lay dormitory.
3. Kitchen. 4. Monks' refectory.
5. Warming house. 6. Cloister.
7. Monks' dormitory. 8. Chapter house. 9. Abbey church.
10. Infirmary kitchen. 11. Infirmary. 12. Infirmary chapel.
13. Cloister.

Buttress

Early English capitol (c. 1220)

Capitol (c. 1230)

DECORATED GOTHIC
A.D. 1275–1375

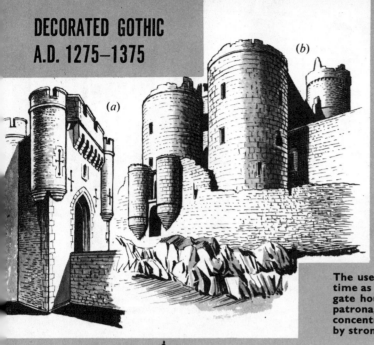

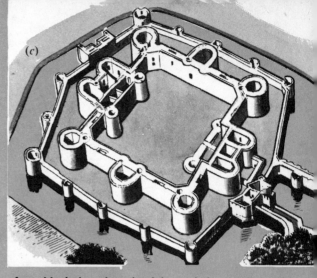

The use of machicolations (openings) began being used at this time as in the illustration (a) of Lewes gatehouse. Strong tower gate houses (b) were built, as at Harlech Castle. Under the patronage of Edward I, castle building saw its finest phase in concentric construction (c). A high inner wall is sub-divided by strong towers, each of which is a strongpoint.

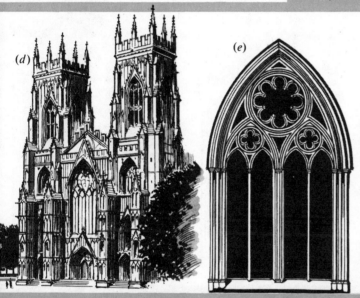

York Minster's west front (d) presents a fine example of this period. Decorated gothic work is generally divided into two parts. The first and earlier period is called geometric and the second and later period is known as curvilinear or flamboyant. These styles show up best mainly in window designs. (e) Early geometric window (c. 1280). (f) Geometric tracery window (c. 1325). (g) Flowing tracery—called flamboyant—style (c. 1360). (h) Decorated gothic arcades. (i) Flying buttress. (j) Foliated capitols. (k) Decorated moulding. (l) Secular living quarters of the upper class. This example is of Markenfield Hall, Yorkshire, of about 1310.

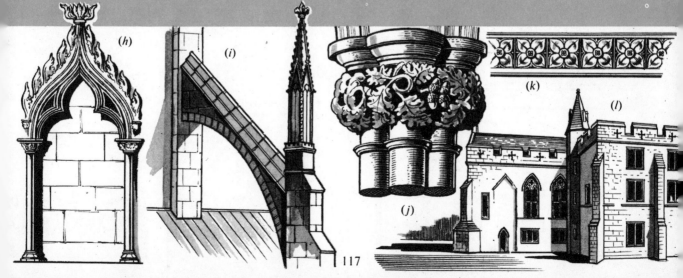

PERPENDICULAR GOTHIC A.D. 1375–1509

(b)

(c)

(a)

(d)

Castle building in England and Wales gradually declined. A few castle-mansions, however, were built in England, and these include Raglan and Hurstmonceux. In Scotland, necessity caused castle building to continue briskly. This gave rise to the development of great tower gatehouses. Examples of these are (a) Doune Castle. The tower-house at Borthwick Castle (b) and others like it showed a return to the keep idea. An illustration of Northern England's robust square construction is Bolton Castle (c). Gun and cannon ports (d).

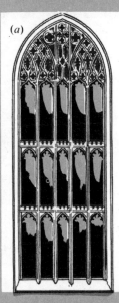

(a)

(b)

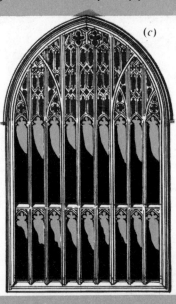

(c)

The Poultry Cross at Salisbury (above) is one of the few market crosses still to be seen in Britain. Another is at Chichester in Sussex. Busy markets were held on the sites they occupied.

Here are further examples of the style. (a) Clerestory window (c. 1500–5). (b) Porch to Gloucester cathedral (c. 1420). (c) Window at Westminster Hall of the late 14th century. (d) Parapet decorative crenellation. (e) Part of a stone, moated manor house, Great Chalfield, Wiltshire (c. 1480). (f) Part of the wonderful fan-vaulting roof at Henry VII's chapel at Westminster Abbey, London. (g) Carved stone finial. (h) Moulded, perpendicular capitol (c. 1395–1450). (i) Delicately carved stone ornamentation.

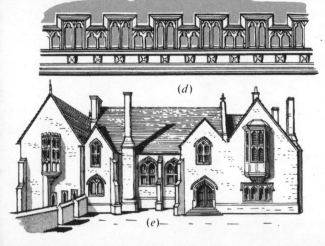

(d)

(e)

(f)

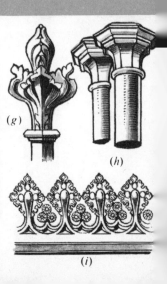

(g)

(h)

(i)

TUDOR
A.D. 1509–1603

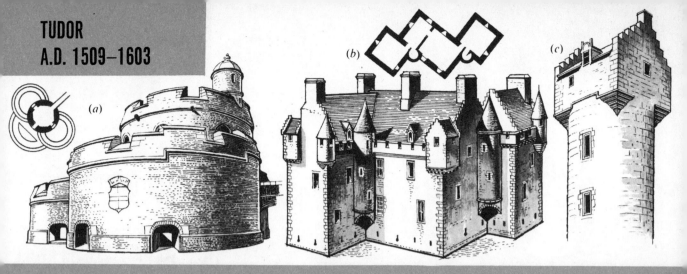

(a)

(b)

(c)

The Tudor age saw the end of the great Gothic era and the birth of the Renaissance age in England. The Tudor age also saw the rise of a new enlightened aristocracy which demanded more comfort and amenities from the buildings they lived in. For the defence of our coasts, many brick gun forts of different designs were built around the south coast by Henry VIII of which St. Mawes in Cornwall (a) is an example. Sixteenth century Scottish castles followed a Z shaped plan (b). Garret chambers (c) were grafted to round drum towers.

(d)

(e)

(f)

(g)

(h)

(i)

(j)

Battlemented gate-houses (d and e) were early Tudor domestic and university features, but were thinly walled. Other examples of the architecture of this period are: (f) Decorated Tudor chimney stacks. (g) Typical Tudor stone cupola. (h) A carved wood corbel bracket, which is typical of this time. (i) Stone parapet cresting. (j) Carved oak panel.

Decorative examples of the domestic, wooden Tudor structure (k) are to be found in Lancashire, Cheshire and Warwickshire. (l) Angle turret and gable end of Cobham Hall, Kent (c. 1597). (m) E-shaped house plan, said to be designed in honour of Elizabeth I. (n) Part of Montacute House, Somerset (1588–1601). (o) Part of a wing of Wallaton Hall, Nottinghamshire, south front (1580–1588).

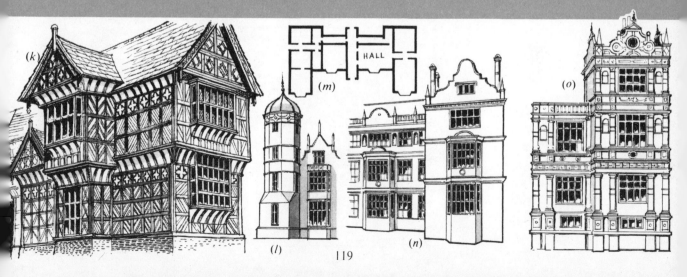

(k)

(m)

HALL

(l)

(n)

(o)

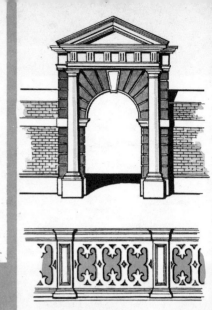

Castle building on a smaller scale continued in Scotland. Castle Frazer, Aberdeenshire (above) is a splendid example. It was built between 1595 and 1636 and has a mock embattled parapet and shows the influence of the Renaissance. Top right: a gateway designed by Inigo Jones. Bottom right: a decorative parapet.

Examples of early Stuart ornamentation are shown here. (a) North front entrance to Hatfield House (c. 1607–12). (b) Member flanking doorway. (c) Detail of twin ornamental columns on single base. (d and e) Brick and stone gables (c. 1629–38). (f) Severe Jacobean façade of Chastleton Hall, Oxon (c. 1603–12). (g) Part of Raynham Hall, Norfolk (c. 1636). (h, i, j) Towers of London churches. (h) St. Stephen Walbrook. (i) St. Edmund the King. (j) St. Martin, Ludgate. (k) Window of about 1663. (l) Finial (c. 1680). (m) Custom House at King's Lynn (c. 1681).

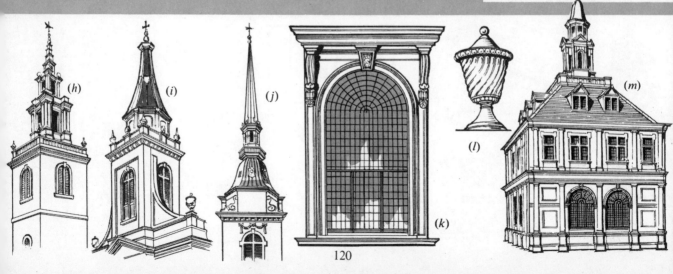

QUEEN ANNE
A.D. 1702–1714

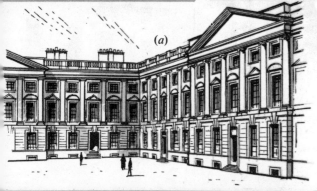

Sir Christopher Wren was the foremost architect at this time, and Hawksmoor and Vanbrugh came to the fore. (a) Part of Christ Church quadrangle, Oxford (c. 1705–11). (b) Façade of St. George's Church, Hanover Square, London.

Above (c) is a typical example of a London Queen Anne terraced house, popular today.

EARLY GEORGIAN
A.D. 1714–1760

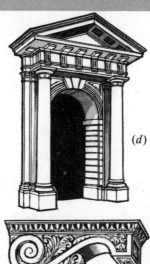

This was the great age of the Palladian school. (a) A window from Horse Guards Parade, London (c. 1751–58). (b) Window of about 1735. (c) Window of about 1726. (d) Doorway of about 1725 with a pediment and Doric style columns. (e) Corbel (c. 1730). (f) The country church at Blandford, Dorset (c. 1735). (g) Part of the front façade of Great Whitley Church, Worcestershire (c. 1735). (h) A supreme example of the Palladian style is Chiswick House, London, built between 1727 and 1736 by Lord Burlington, one of the aristocratic leaders of the movement. It is a perfectly balanced building from all angles. (i) Part of the south front of Lyme Park, Cheshire (c. 1720–26) for which the architect was Giacomo Leoni, a leading exponent of this style.

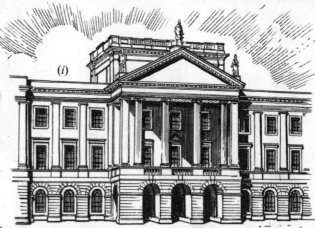

LATE GEORGIAN
A.D. 1760–1800

(a)

(b)

(c)

(d)

The great architects of this period included Robert Adam and Sir William Chambers. (a) Gateway at Blenheim Palace, Oxon (Sir William Chambers). (b) South front of Kedleston Hall, Derbyshire, started c. 1761 (Robert Adam). (c) Part of the plaster ceiling ornamentation at Paxton House, Scotland (Adam brothers). (d) Provincial town architecture, Fairfax House, York (c. 1770). Robert Adam and his three brothers were distinguished architects, whose work contrasted with that of Sir William Chambers.

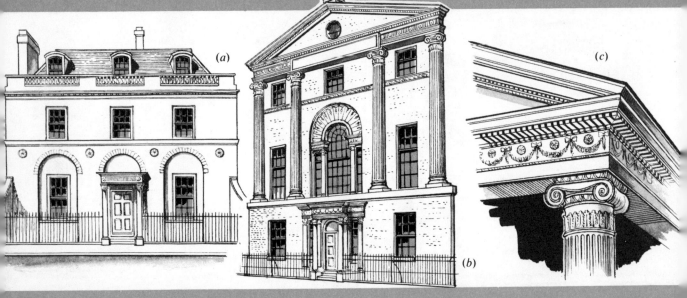

(a)

(b)

(c)

Above: (a) Smaller London suburban house, c. 1775. (b) The splendid façade of the Royal Society of Arts in London (c. 1772–4) designed by Robert Adam. (c) A capitol, column, pediment and entablature of the Ionic Order designed by Robert Adam for Kenwood House, Hampstead, in about 1770. Below: (a) Window, c. 1780. (b) Window, c. 1797 by John Nash. (c) Window of 1778–86 by Sir William Chambers. (d) Adam style decoration. (e) Garden finial of the late 18th century. (f) Porch at Chandos House, Marylebone (c. 1770) by Robert Adam. The Adam brothers, notably Robert, dominated the architecture of this time. When Robert was designing a house, he took great care with every detail, even down to the actual keys used. Syon and Kenwood Houses in London are among his work.

(a)

(b)

(c)

(d)

(e)

(f)

REGENCY & VICTORIAN
A.D. 1800–1900

(a)

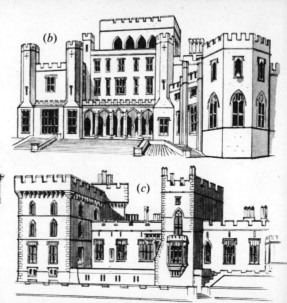

(b)

(c)

Many styles appeared in this period. (a) The Royal Pavilion at Brighton designed by John Nash, c. 1815–1820. (b) Ashridge Park, Herts., begun in c. 1806 by James Wyatt. (c) Extension to Windsor Castle begun in 1824 by Sir Jeffry Wyattville.

(a)

(b)

(c)

(d)

Above: (a) Entry arch to Euston railway station, London. (b) Big Ben at the Houses of Parliament, London, designed by Sir Charles Barry and A. W. N. Pugin, 1836–1865. (c) Bow-fronted terraced houses at Hove, Sussex (c. 1826). (d) Tower entrance to University Museum, Oxford, 1855–60. Below: (e) Capitol from the Natural History Museum, London, demonstrating the Victorian over-enthusiasm for the Gothic style, which misses the true feeling of that age. (f) Victorian domestic building. This is the Red House, Bexleyheath, Kent, once the home of William Morris, a poet, artist, decorator, manufacturer, printer and socialist. It was built in 1859. (g) A brick terraced house built in the second half of the 19th century. Building was encouraged by the industrial revolution at this time and the coming of the railways.

(e)

(f)

(g)

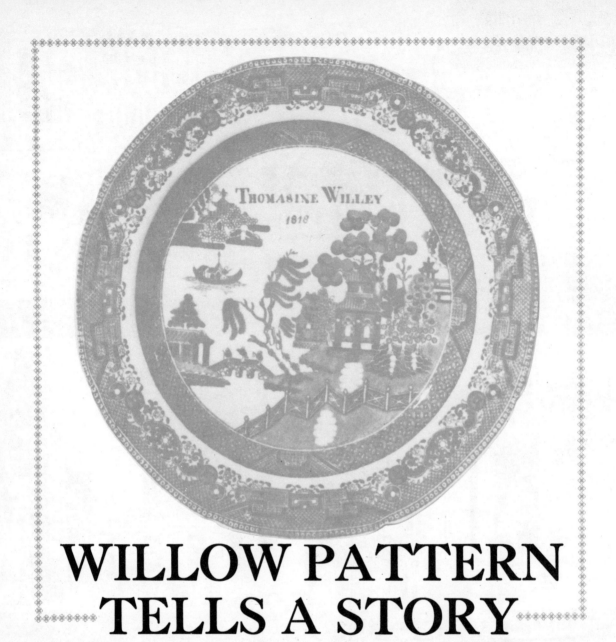

WILLOW PATTERN
TELLS A STORY

ONE of the ever-popular designs for table-ware is the blue-and-white 'Willow Pattern.' The figures upon it, bridge, house, river, doves and trees tell a story whose design is now standard. But many old plates show that there were once variations.

The boat in which the young lover sails to seek the girl who is locked up in the tower is shown in different positions. Sometimes there are only the girl and the boy upon the bridge, but the standard version shows the girl's father following. The number of apples upon the fruit-tree varies, and the doves, who are the lovers changed into birds to escape the father, also vary in position.

The Caughley Pottery in Shropshire was the first factory to introduce the 'Willow Pattern' in the year 1780. It is thought that Thomas Turner, owner of Caughley at the time, thought up the idea after a visit to Paris in search of new designs.

The 'Willow Pattern' which resulted is meant to appear Eastern, but is wholly Western in its legendary sentiment.

A Caughley apprentice at the time was Thomas Minton, and it was he who engraved the copper-plate for the 'Transfer-Printing.'

'Transfer-Printing' was an English way of decorating china. Special ceramic ink was placed over the plate and wiped, leaving the ink in the grooves and cross-hatchings. Then a thin paper was placed on the copper-plate and lifted. While the ink on the paper was still wet it was pressed on to the porcelain, leaving the inky design.

In 1797 Thomas Minton set up his own pottery factory at Stoke-on-Trent, and took his copper-plate of the 'Willow-Pattern' with him. This is still in existence, but it is thought that he may have made and sold differing versions to other potteries before he set up on his own.

A Wet Sheet and a Flowing Sea

This famous poem comes from "The Songs of Scotland, Ancient and Modern" by Allan Cunningham who lived from 1784 to 1842. After working as a stone mason, he became the secretary to a sculptor. Poetry entered his life when he supplied material for R. H. Cromeks "Remains of Nithsdale and Galloway Song." He published "Traditional Tales of the English and Scottish Peasantry" in 1822, and his "Songs of Scotland" appeared in 1825. Many of his short pieces and imitations of ancient ballads became very popular.

A wet sheet and a flowing sea,
　A wind that follows fast
And fills the white and rustling sail
　And bends the gallant mast;
And bends the gallant mast, my boys,
　While like the eagle free
Away the good ship flies and leaves
　Old England on the lee.

O for a soft and gentle wind!
　I heard a fair one cry;
But give to me the snoring breeze
　And white waves heaving high;
And white waves heaving high, my lads,
　The good ship tight and free—
The world of waters is our home,
　And merry men are we.

There's tempest in yon hornéd moon,
　And lightning in yon cloud;
But hark the music, mariners!
　The wind is piping loud;
The wind is piping loud, my boys,
　The lightning flashes free—
While the hollow oak our palace is,
　Our heritage the sea.

A. Cunningham

JOIN A CHOIR— AND SEE THE WORLD!

HOW can you travel if you cannot afford the fare? To become a sailor was once every boy's idea of a gateway to adventure. But your voice can also be a ticket to a tour of exciting places if you really love music and can sing and play better than most trained young musicians.

But there is something else! You must belong to what many people consider to be the finest choir in Europe. This is the choir of St. John's College in Cambridge.

Not only has it sung on the continent, but in the summer of 1970 it was booked to go on a three-week tour of the United States and Canada. Travel is a bonus for the boys and a treat for the people who hear them sing. Montreal, Buffalo City, Saratoga Springs, Tanglewood, Hartford (Connecticut) and New York . . . these were the places that were to become no longer names on a list but the source of vivid memories.

Normally, such a tour would have cost each boy a lot of money. But fees paid by Canadian and American organisations, a recording company and the Oxford University Press covered all the expenses.

The choir's voices carry them through Britain, too. Recitals have been given in cathedrals and halls in England and Wales and more are to come. The hard work which these tours demand is enjoyed by the keen choristers, who also have chances to amuse themselves. When there is time for some fun, such as a swim in the sea or a mountain climb, they are quick to take the opportunity.

These tours break the strict routine of their lives. Daily practice and sung services five days a week and two on Sundays demand a high level of self-discipline. But to a boy who loves music, this is not a hardship.

Singing in a choir, like playing cricket or football, depends for its success upon team work, and at St. John's each boy quickly learns to give of his very best for the benefit of the choir.

Mr. George Guest, the choir master, extracts from his 28 voice choir of 12 undergraduates and 16 other choristers not only good singing but a deeply felt love of music.

A great deal of time is spent learning music, playing instruments and singing, and because the boys are aged between eight and 13 years they have to work extra hard at their lessons.

The choir's excellence has grown slowly. Fourteen years ago, St. John's College School had only about 50 day pupils who lived locally and the boys who formed the choir were drawn from these. Then, in 1956, the school moved to new buildings with a different headmaster and the number of pupils rose rapidly to 100 day boys and 40 boarders.

The choir master decided to improve the standard of singing and advertised "voice trials" for places in the choir for probationers who could after-

**Above: the choir sings at choral Evensong for a TV broadcast from a Roman Catholic church in Holland.
Left: the choir of St. John's College, Cambridge, are given advice before a service.**

wards become choristers if they proved to be good enough. So many good singers came along, that the improvement was dramatic.

Eventually, the choir has become so good and the competition to join it so keen that, nowadays, as many as 40 boys apply for the four or so vacancies which occur each year.

However, this keen competition should not deter any boy with a musical talent from applying for a place in this choir or in any of the 30 or so others maintained by cathedrals and colleges in Britain. All welcome applications from talented youngsters between the ages of 7½ to 9½ years.

Those who succeed in being accepted for St. John's choir become boarding pupils at the college school. The scholarship which they are awarded reduces their school fees by from £200 to £300 a year, and each boy receives a first class preparatory school education.

SCHOLARSHIPS

When a boy reaches the age of 13, he may be able to go on to a public school with a strong musical tradition. Many offer music scholarships and some have special choristers' places worth between £100 and £300 a year.

By the time the boy enters his public school, he will have reached a high standard of proficiency in playing the piano and a second instrument, as well as having a knowledge of music theory.

Being a chorister does not, as is often believed, mean a restricted life with little freedom. Hobbies and special interests absorb the boys' spare time, although organised games are limited.

But there are diversions, even during choir practice. Recently, a few pigeons fluttered into St. John's College Chapel in Cambridge and their flying distracted the choristers and choral students from their rehearsal.

Getting rid of them was no problem for the choir master.

He shouted to the accompanist in the organ loft above the choir, "Give them a blast on the trompetta." As the trompetta stop on the organ was pressed, a resounding blast of trumpet sounds echoed through the chapel and scared the pigeons away.

When the last bird had gone, the choir master raised his hand as a signal for the rehearsal to resume. The organ began its introduction and 28 sweet voices in perfect harmony filled the air.

There will be more rehearsals and more services and concerts. Eventually, the choristers will leave their public school to make their way in the world. Many will become famous like Palestrina, who composed religious music, Bach, Dr. Burney and Sir Arthur Sullivan and the hundreds of other eminent musicians who obtained their first instruction as choristers.

Not only will they have joined a choir and seen the world, they will have made their mark in it as well.

The Arts Picture Quiz

Now that you have read the Arts section, see if you can answer this simple quiz.
The answers are at the back.

 1 Which author of exciting books was called "a sailor with a fiery pen"?

 2 Of what offence was the "fiery pen" author accused when he appeared before a tribunal?

 3 Which country might you visit to see this actor? How long could a play last there?

 4 Who painted this portrait of himself and another which fooled a king?

 5 The woman in this picture became the Queen of England? Do you know her name?

 6 What is this strange-looking structure called and where is it to be seen?

 7 The entrance to this castle has a distinctive feature. What is it called?

 8 Can you recognise the period which this house belongs to and what years does it cover?

 9 Can you name the style of building shown here and give the name of this house?

 10 This kind of Norman fortress has a special name. Do you know what it is?

 11 This choir is world famous and travels a great deal. What is its name?

12 If you sang in this choir, which countries might you have visited in 1970?

BE A SPORT!

Do not read these answers until you have looked at the quiz questions on pages 24, 48, 80, 104 and 128.

HISTORY QUIZ

(page 24)

1. Athens. 1896.
2. Dorando Pietri. He was disqualified because he was helped by others.
3. Jesse Owens. Four gold medals.
4. A galleon. "Vasa". It sank.
5. Fridtjof Nansen.
6. The Eskimos.
7. George Villiers.
8. Narvik.
9. Eire.
10. Cyrillic. 32 letters.
11. Kosher. Right to left.
12. Ampersand. And.

SCIENCE QUIZ

(page 48)

1. In 1783. Hot air.
2. Wilbur and Orville Wright. 852 ft. (59 seconds).
3. Germany. In May, 1941.
4. London Bridge.
5. Sydney Harbour Bridge. New South Wales, Australia.
6. The Menai Strait Bridge.
7. By sliding a free panel in the lid.
8. A hidden steel rod.
9. Water.
10. The Dead Sea. The density of its salt content.
11. Cuprum.
12. About five million tons.

GEOGRAPHY QUIZ

(page 80)

1. North Island. New Zealand. Rotorua.
2. Cyprus. The Egyptians.
3. Bought by Britain in 1878. It became independent in 1960.
4. The Western and Eastern Sierra Madres.
5. Stockholm City Hall. Venice of the North.
6. Between 25,000 and 35,000.
7. National Romantic Style.
8. Hematite iron ore.
9. Lang Hancock. Tom Price. Western Australia.
10. Jamaica.
11. To send messages.
12. Oil drilling. Alaska. Freezing fog.

NATURE QUIZ

(page 104)

1. Hedgehog.
2. Halibut. Almost 8ft.
3. A clamp.
4. A stalk of seeds.
5. An opossum.
6. Australia. A kangaroo.
7. A grass cricket.
8. 130 beats a second. A bumble bee.
9. The Blue Whale. The Antarctic.
10. India and Africa. Alexander the Great.

ARTS QUIZ

(page 128)

1. Jack London.
2. Of slapping a servant.
3. Japan. All day.
4. Hans Holbein the Younger.
5. Anne of Cleves.
6. The Poultry Cross, Salisbury.
7. Double-towered.
8. Tudor; 1509-1603.
9. Palladian. Chiswick House.
10. A shell keep.
11. The choir of St. John's College, Cambridge.
12. Canada and the U.S.A.